# Intermediate 1

# PHYSICS

**Arthur Baillie**
**Andrew McCormick**

The Publishers would like to thank the following for permission to reproduce copyright material:
Photo credits: © Erik Freeland/Corbis; © Hulton-Deutsch Collection/Corbis; Pure Digital – Imagination Technologies; Science Photo Library; © Bettmann/Corbis; © Chinch Gryniewicz, Ecoscene/Corbis; Volker Steger/Science Photo Library; Tony Buxton/Science Photo Library; © Mark E. Gibson/Corbis; © Lucidio Studio Inc./Corbis; © Ralph White/Corbis; D. Roberts/Science Photo Library; © Toshiyuki Aizawa/Reuters/Corbis; Sheila Terry/Science Photo Library; Gary Parker/Science Photo Library; © Lawrence Manning/Corbis; Gusto/Science Photo Library; Western Ophthalmic Hospital/Science Photo Library; Bsip, Laurent/Science Photo Library; Will & Deni Mcintyre/Science Photo Library; Nih/Custom Medical Stock Photo/Science Photo Library; © Jeff Albertson/Corbis; David M. Martin, M.D./Science Photo Library; © Royalty-Free/Corbis; John Mclean/Science Photo Library; Matt Meadows/Peter Arnold Inc./Science Photo Library; Cnri/Science Photo Library; Martyn F. Chillmaid/Science Photo Library; Colin Cuthbert/Science Photo Library; Alexander Tsiaras/Science Photo Library; © Lester V. Bergman/Corbis; Simon Fraser/Ncct, Freeman Trust, Newcastle-Upon-Tyne/Science Photo Library; Josh Sher/Science Photo Library; © Corbis Sygma; Dr Ray Clark & Mervyn Goff/Science Photo, Library; © Ekbersader Bikem/Corbis Sygma; Simon Fraser, Hexham General Hospital/Science Photo Library; Pascal Goetgheluck/Science Photo Library; Michel Viard, Peter Arnold Inc./Science Photo Library; James Stevenson/Science Photo Library; Maximilian Stock Ltd/Science Photo Library; © Pete Stone/Corbis; © Tim Kiusalaas/Corbis; David Nunuk/Science Photo Library; © Dave Bartruff/Corbis; © Helen King/Corbis; Takeshi Takahara/Science Photo Library; Toyo Tyre Uk; © Reuters/Corbis; © David Stoecklein/Corbis; © Joe Mcbride/Corbis; © William Whitehurst/Corbis; Adam Hart-Davis/Science Photo Library; © Martyn Goddard/Corbis; Trl Ltd/Science Photo Library; © Jim Cummins/Corbis; © Mark A. Johnson/Corbis; © Reuters/Corbis; © Paul Almasy/Corbis; © Patrik Giardino/Corbis; © Roy Morsch/Corbis; © Dimitri Iundt; Tempsport/Corbis; © Michael Cole/Corbis; © Patrik Giardino/Corbis.

Acknowledgments: Page make-up and additional illustrations by Hardlines Ltd, Oxford.
Every effort has been made to trace all copyright holders, but if any have been inadvertently overlooked the Publishers will be pleased to make the necessary arrangements at the first opportunity.

Whilst every effort has been made to check the instructions of practical work in this book, it is still the duty and legal obligation of schools to carry out their own risk assessments.

Although every effort has been made to ensure that website addresses are correct at time of going to press, Hodder Gibson cannot be held responsible for the content of any website mentioned in this book. It is sometimes possible to find a relocated web page by typing in the address of the home page for a website in the URL window of your browser.

**The front cover shows streamers of light inside a plasma globe, which is a glass sphere filled with gas at low pressure. At its centre is a metal ball that is charged with electricity. When the voltage is high enough, the electricity is emitted across the sphere to the glass wall. The gas becomes ionised along the path which the electricity follows, causing it to give off light.**

Orders: please contact Bookpoint Ltd, 130 Milton Park, Abingdon, Oxon OX14 4SB. Telephone: (44) 01235 827720; Fax: (44) 01235 400454. Lines are open from 9.00 – 5.00, Monday to Saturday, with a 24 hour message answering service. Visit our website at www.hoddereducation.co.uk.
Hodder Gibson can be contacted direct on: Telephone: 0141 848 1609; Fax: 0141 889 6315; email: hoddergibson@hodder.co.uk.


First published in 2004 by
Hodder Gibson, an imprint of Hodder Education, an Hachette UK Company
2a Christie Street
Paisley PA1 1NB

Impression number 10 9 8 7 6 5
Year 2010



Cover photo by Alfred Pasieka/Science Photo Library (T194/550)

Typeset in Utopia 8pt by Hardlines Ltd, Charlbury, Oxford
Printed in Dubai for Hodder Gibson, 2a Christie Street, Paisley, PA1 1NB, UK
A catalogue record for this title is available from the British Library



ISBN 13: 978 0 340 81700 1

# Contents

Unit 1 Telecommunications 1

1 **Radio** 2
2 **Television** 7
3 **Satellites** 11
4 **Optical fibres** 16
5 **Telephones** 20

Unit 2 Practical Electricity 27

1 **Electrical circuits** 28
2 **Resistance** 39
3 **Mains electricity** 43

Unit 3 Radiations 57

1 **Lasers** 58
2 **X-rays** 70
3 **Gamma rays** 74
4 **Infrared and ultraviolet** 78

Unit 4 Sound and music 85

1 **Sound waves** 86
2 **Speed of sound** 92
3 **Using sound** 96
4 **Amplified sound** 101

Unit 5 Movement 109

1 **Forces** 110
2 **Speed and acceleration** 123
3 **Moving objects** 129

Unit 6 Electronics 139

1 **Input, process and output** 140
2 **Digital logic gates** 148

Unit 1

# Telecommunications

**Every day, people send and receive lots of information. This could be from TV or radio or the internet or using mobile or fixed line telephones. Some long distance systems use satellites. All of these need a transmitter and a receiver. This topic will look at how these systems work.**

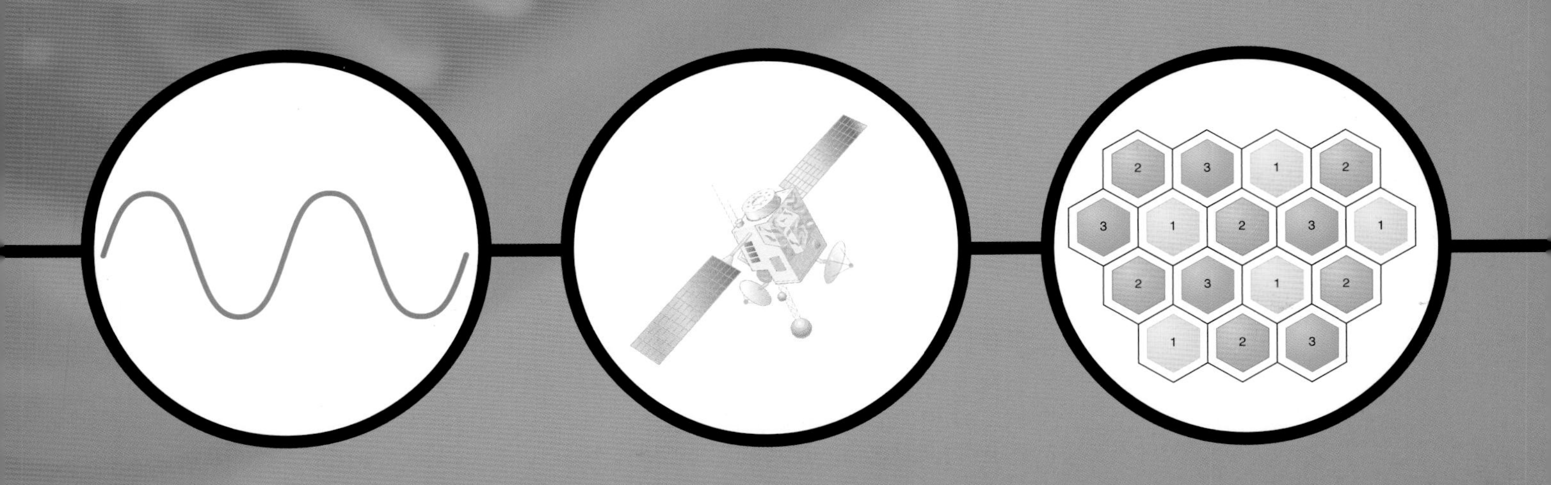

# Radio

## After studying this section you should be able to...

1 State that radio communication does not require wires between the transmitter and receiver.
2 State that radio signals are waves which transfer energy.
3 State that radio signals are sent through air at a speed of 300 000 000 metres per second.
4 Complete a block diagram of a radio receiver in the correct order.
5 Describe the function of the aerial, tuner, amplifier and loudspeaker.
6 State the frequency of a radio signal is the number of waves per second.
7 State that frequency is measured in hertz.
8 State that a radio station can be identified by the frequency of the signal that it transmits.

**Figure 1.1** *Early Marconi transmitter.*

John was using his mobile phone to talk to his friends. The mobile phone also had a radio built into it. The quality of sound was not as good as his radio at home. There he had a digital audio radio. It allowed him to listen to a large number of stations and to hear them with no hisses and crackles. How did the radio signals reach this small radio in his mobile phone when there were no wires?

The first radio signals were sent about 1879 by a German called Heinrich Hertz. But it wasn't until the early 1900s that Marconi managed to send radio signals over a large distance and eventually across the Atlantic from England to Newfoundland in Canada. Since then radios have developed on their own, as part of hi-fi equipment, and even on the internet and in mobile phones.

## How does a radio work?

All communication systems use a transmitter and receiver.
A transmitter sends out the signals from the radio station and the receiver collects all the signals from all the radio stations.
In figure 1.1 you can see a radio transmitter.

- Radio is an example of very long-range communication. Walkie talkies can only allow messages to travel a short distance.
- Radio does not need wires between the transmitter and the receiver.
- Radio signals travel very quickly. Their speed in air is 300 000 000 metres per second.
- The signals travel as waves, like water waves, and so carry energy.

**Figure 1.2** *Sattelite dishes at a major event.*

The very fast speed of radio signals means that we can see or hear events almost as they happen, so live broadcasts are received almost at the instant they happen anywhere in the world. The distance from Glasgow to New York City is 5200 000 metres, and to send a TV signal would take about 0.017 seconds. This time is so small that we would not notice it and we would see events almost as they happened. The Olympics and the World Cup can be seen live (figure 1.2).

Signals can be sent out with different frequencies.

Different radio stations transmit on different frequencies. Frequency is the number of waves per second.

Frequency is measured in hertz (Hz). This means that 5 Hz is 5 waves per second.

### Examples

20 waves pass a point in 5 seconds. What is the frequency of the wave?

Solution

$$\text{Frequency} = \frac{\text{number of waves}}{\text{time}}$$

$$= \frac{20}{5}$$

$$= 4\,\text{Hz}$$

## Radio reception

The different parts of the radio receiver are shown in figure 1.3.

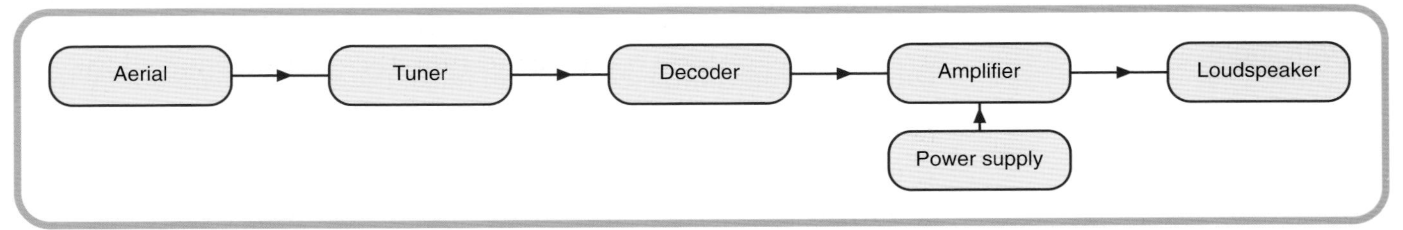


**Figure 1.3**

### The aerial

The aerial receives *all* of the radio signals and changes them into electrical signals.

## Tuner

A radio is able to receive radio waves from many different stations. To keep the signals separate, each station transmits radio waves at a different frequency. The tuner lets you select the particular radio station you want to receive.

## The decoder

This part is not required to be known in this course.

There are two parts to the received signal. One part is required to allow the waves to travel large distances. This part is not needed when the signal has reached the radio. The other part is the bit that we need, called the audio wave. The decoder separates the two parts.

## The amplifier

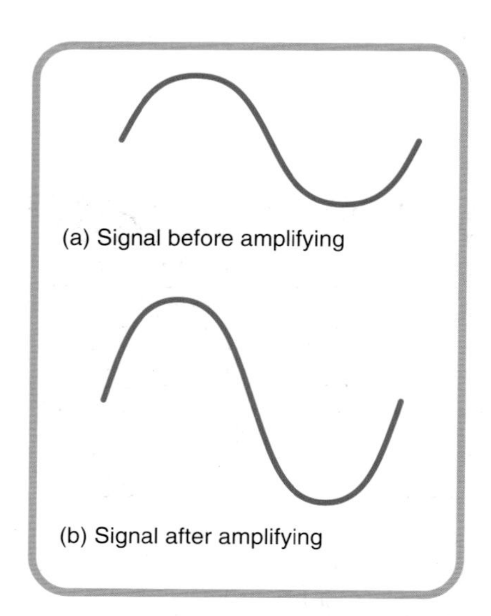


**Figure 1.4** *Electrical signal in a radio before and after amplification.*

The signal that is left after passing through the decoder has very little energy. This signal has to be amplified. The amplifier increases the strength of the received signal. It makes the energy of the received signal larger. To do this the amplifier needs to be supplied with energy from a battery or power supply.

Figure 1.4 shows the electrical signal in a radio before and after amplification. The amplifier has made the height of the signal bigger and so it now has more energy. This extra energy is supplied from the battery or electrical supply connected to the amplifier.

Check that the amplifier makes the signal bigger by connecting a signal generator to an amplifier and then an oscilloscope. Look at the pictures on the oscilloscope screen without the amplifier operating and then with it switched on. Notice that it is only the height of the signal that increases, the frequency does not change. That is, there is always the same number of waves on the screen.

On a radio the volume control can make the energy given out by the radio larger.

## Listening

The amplified electrical signal must now be turned into sound energy so you can hear it. This requires a loudspeaker.

The loudspeaker changes electrical energy into sound energy.

## The radio band

Table 1.1 shows the radio frequency bands in the UK.

| Radio waveband | Use and example(s) |
|---|---|
| Low frequency (long wave) | Medium to long distance. AM band. Radio 4 (200 kHz). Wavelengths around 1000–2000 m. |
| Medium frequency (medium wave) | Both local and distant sound broadcasts. AM radio. National radio stations such as Radio 5. Some local radio stations. |
| High frequency (short wave) | Long distance communication, ship-to-shore links, navigation, radio beacons and amateur radio. |
| Very high frequency (VHF) | Short distance communication, FM sound broadcasts, often in stereo. Local radio stations. |
| Ultra high frequency (UHF) | TV channels, mobile phones and police. |
| Super high frequency (SHF) | Microwave links, radar, satellite TV and international phone links. |

**Table 1.1**

## Radio stations

Table 1.2 gives a list of some radio stations, which can be found in many newspapers.

| Station name | Frequency |
|---|---|
| Clyde 1 | 102.5 MHz (102 500 000 hertz) |
| Radio Scotland | 810 kHz (810 000 hertz) |
| Radio 1 | 99.8 MHz (99 800 000 hertz) |
| Radio 2 | 90.2 MHz (90 200 000 hertz) |
| Radio 5 | 909 kHz (909 000 hertz) |
| Virgin Radio | 1215 kHz (1 215 000 hertz) |

**Table 1.2**

### Physics for you

Digital radio is the latest development. It is sometimes called DABS (Digital Audio Broadcasting System). The receiver is expensive, but has certain advantages (figure 1.5):

- In some areas there is interference to the signals caused by the signals reflecting off buildings and hills. Using a special frequency technology this is eliminated and the same frequency can be used across the country. This means that you do not need to retune your set as you move to another area.
- The sound quality is noticeably better.
- There are a wider range of stations. You are able to listen to about 350 stations. Some will be old favourites in digital, but there are also stations for interests like soul or jazz music.

**Figure 1.5**

## Key Points

- Radio communication does not require wires between the transmitter and the receiver.
- Radio signals are waves which transfer energy.
- Radio signals are sent through air at a speed of 300 000 000 metres per second.
- The key components of a radio are the aerial, tuner, amplifier and loudspeaker.
- The frequency of a radio signal is the number of waves per second.
- Frequency is measured in hertz.
- A radio station can be identified by the frequency of the signal that it transmits.

## Section Questions

**1** Copy and complete the following sentences using the appropriate words from the list: energy, receiver, wires, 300 000 000, transmitter, waves.
In a radio system the signals are sent out by a ______ and received by a ______. There is no need for any ______ between these two parts. The signals are ______ which send out ______. They are sent through the air at a speed of ______metres per second.

**2** The frequency of a radio station is 98 000 hertz. What does this mean?

**3** What is the purpose of the tuner in a radio?

**4** a) What does an amplifier do to the electrical signal in a radio?
b) What does the amplifier NOT change in the electrical signal?

**5** Table 1.3 contains information about radio stations and their frequencies.

| Radio Station | Frequency in hertz |
|---|---|
| Radio Scotland | 810 000 |
| Radio 1 | 998 000 |
| Classic FM | 101 000 000 |
| Talk Sport | 1089 000 |

**Table 1.3**

a) Which frequency does Classic FM use?
b) Which frequency does talk sport use?

**6** The parts of an old radio are found in a drawer. They are not connected together. The parts are loudspeaker, tuner, aerial, amplifier, decoder and battery. Put these parts in the correct order.

**7** Water waves are sent across a pool. Steve counts them and measures 30 waves in 6 seconds. What is the frequency of the wave?

**8** A battery operated radio can be placed in different rooms of a house and it still works. How does the signal get from the transmitter to the radio?

**9** What is the purpose of the aerial in a radio?

**10** What useful energy change takes place in a loudspeaker?

# Television

## After studying this section you should be able to...

1. State that TV signals are radio signals but at a higher frequency.
2. State that a TV station can be identified by the frequency of its signal.
3. Complete a block diagram of a TV receiver showing in order the aerial, tuner, decoders, amplifiers, tube and loudspeaker.
4. Describe the function of the aerial, tuner, amplifiers, tube and loudspeaker.
5. State that mixing red, green and blue light produces all the colours seen on a TV set.

### The television receiver

A television transmitter sends out TV signals. These signals are similar to radio signals but have a higher frequency. Different TV stations send out signals with different frequencies.

Televisions seem to have been around for a long time, but the first pictures were shown by John Logie Baird in only 1924. This used a mechanical system to scan the picture seen by the camera (figure 1.6). Electronic scanning was developed in America. The BBC started broadcasting in 1936, and until 1955 there was only one channel. Early sets had black and white screens only. Nowadays we can have LCD (liquid crystal display) screens, which are very thin, or large projection screens or even plasma ones. A plasma screen is shown in figure 1.7.

**Figure 1.6** *Early John Logie Baird system.*

**Figure 1.7** *A plasma screen TV.*

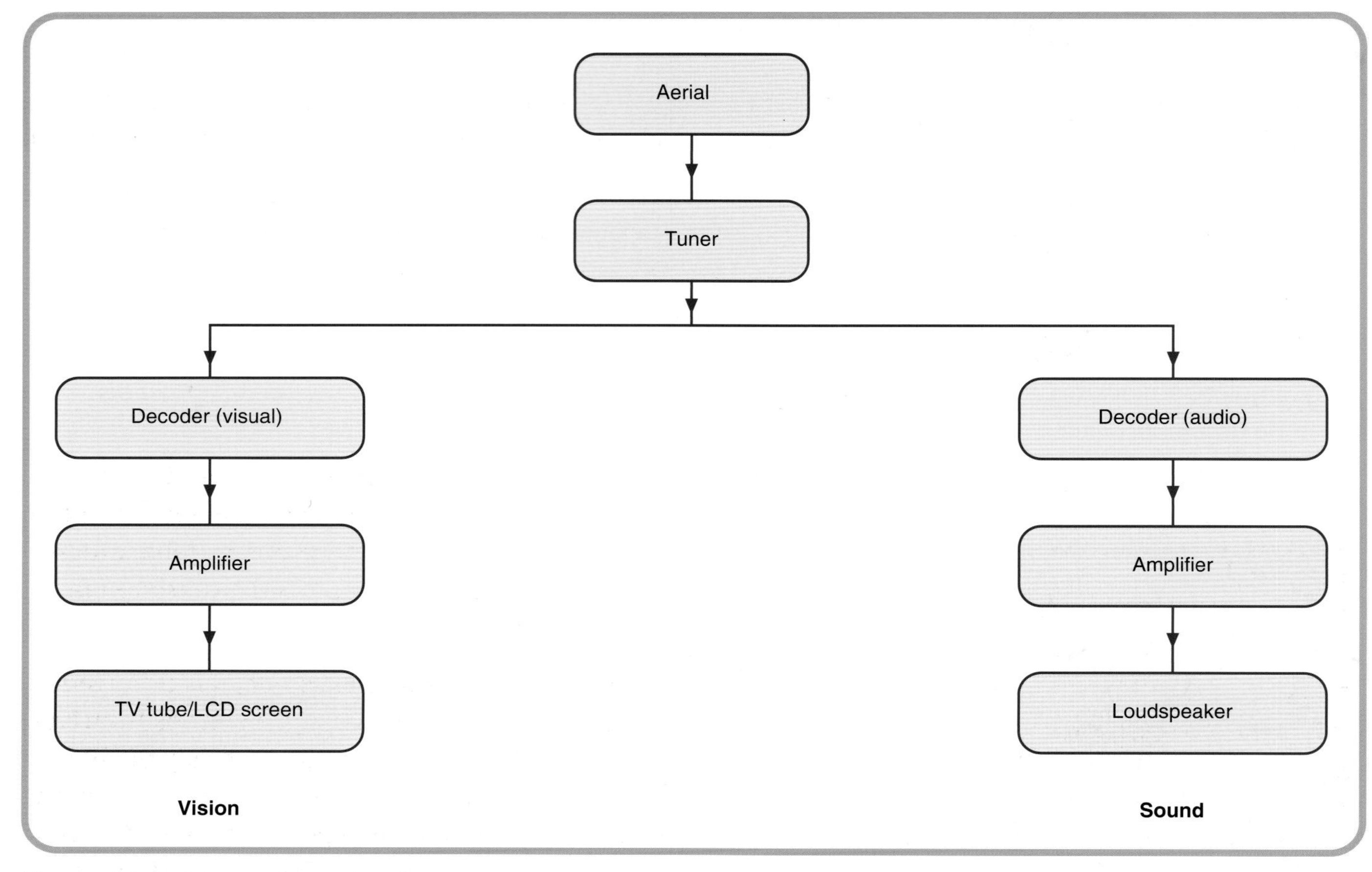


**Figure 1.8** *The main parts of a TV.*

The main parts of a TV are detailed below and shown in figure 1.8:

- Aerial — picks up all of the TV signals and changes them into electrical signals.
- Tuner — picks out the frequency of the TV station you want.
- Decoder (vision) — selects the picture signal from the waves.
- Amplifier (vision) — makes the picture signal stronger.
- TV tube — turns the electrical signal into light.
- Decoder (audio) — selects the sound signal from the waves.
- Amplifier (audio) — makes the sound signal stronger.
- Loudspeaker — turns the electrical signal into sound.

The signal is split into two parts: the audio or sound part and the picture or visual parts.

The audio and visual signals are sent out separately by the transmitter. This sometimes means that although we can see a programme, we cannot hear the sound due to a fault in the transmission.

A battery or mains electricity is required to make the television receiver operate.

The main energy change in the loudspeaker is electrical to sound. The main energy change in the TV screen is electrical to light.

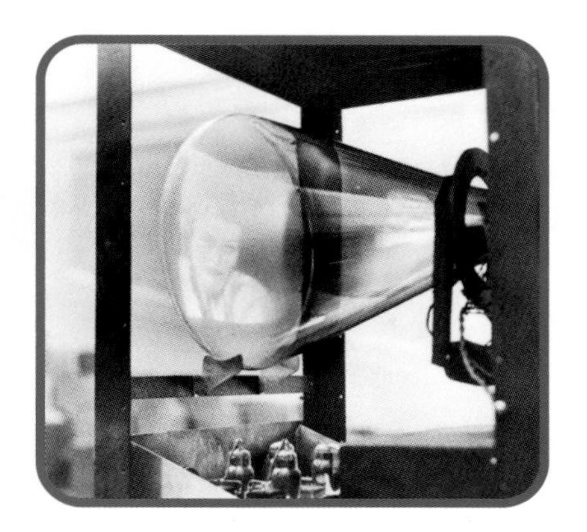

Figure 1.9 *A TV tube.*

## The picture tube

The electron gun, which is at the narrow end of the tube, produces negative charges called electrons and 'fires' them as an invisible beam.

As the electrons cannot travel very far in air, there is a vacuum (no air) inside the tube to allow them to reach the screen. A special coating on the screen produces a tiny spot of light when hit by the electrons (figure 1.9).

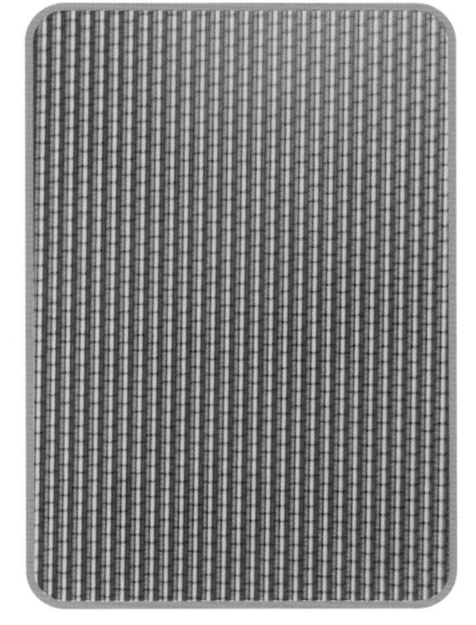

**Figure 1.10** *Colour composition of a television screen.*

## The colour TV tube

A colour TV has three electron guns and a screen coated with about one million tiny dots arranged in triangles. When these dots are hit by electrons, one dot in each triangle gives out red light, another green light and the third blue light. It is the material that the dots are made of which gives out colour (figure 1.10).

As the three electron beams sweep or scan the screen, it is arranged that each beam strikes only dots of one colour, e.g. electrons from the 'red' gun only hit 'red' dots. When a triangle is struck it may be that the red and green electron beams are very strong (intense), but not the blue. The triangle will give out red and green light strongly and appear yellowish. The triangles are struck in turn, and since the dots are so small and the sweeping so fast, we see a continuous colour picture.

The mixing of the three different colours can produce other colours (figure 1.11).

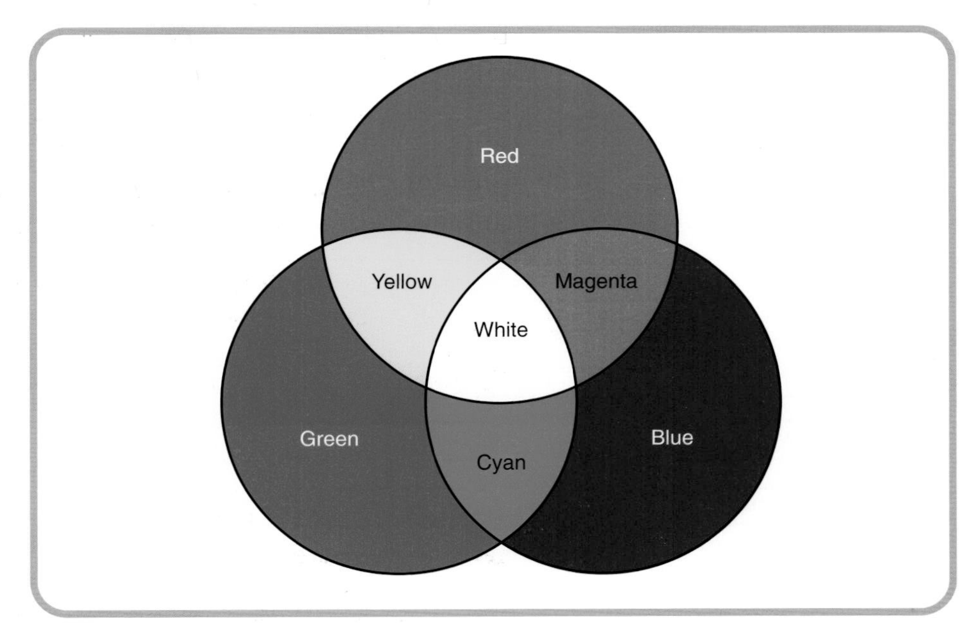


**Figure 1.11**

## Physics for you

LCD screens use special crystals which light up when an electrical signal operates them. Each little part of the colour screen is called a pixel.

The screen is very thin compared to an ordinary TV but you cannot view it from the side.

There are new screens being developed, such as Organic Light Emitting Diodes (OLED). These will be thinner and more reliable than LCDs (figure1.12)

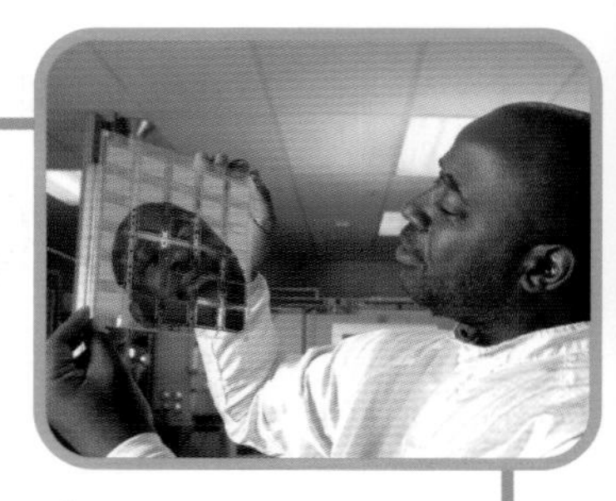

**Figure 1.12**

## Key Points

- TV signals are radio signals but at a higher frequency.
- A TV station can be identified by the frequency of its signal.
- A block diagram of a TV receiver shows in order the aerial, tuner, decoders, amplifiers, tube and loudspeaker.
- The aerial, tuner, amplifiers, tube and loudspeaker have different tasks in a TV.
- Mixing red, green and blue light produces all the colours seen on a TV set.

## Section Questions

**1** How does the frequency of a radio signal compare with that of TV signal?

**2** When you press a button on the remote control, more electrons strike the screen on the tube. How will you notice this on the screen?

**3** What is the purpose of the tube in a TV set?

**4** Which three colours can be mixed together to give all the colours in a TV picture?

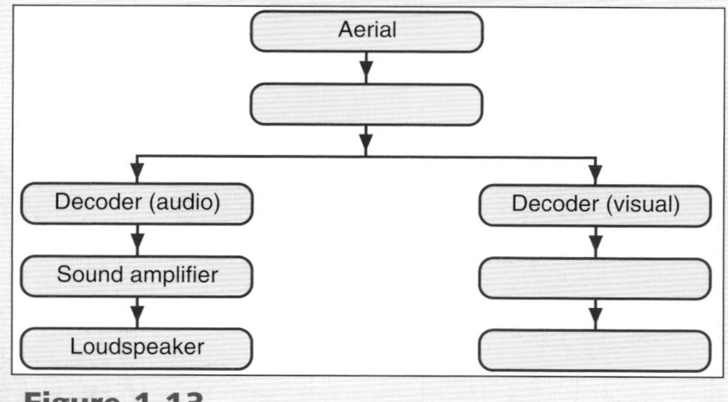


**Figure 1.13**

**5** Figure 1.13 shows parts of a TV. Complete the missing parts of the diagram.

**6** Why are there two sets of amplifiers in a TV?

**7** The frequency of BBC1 from one transmitter is 630 000 000 Hz and the frequency of BBC2 is 671 000 000 Hz. Which station operates at the higher frequency?

**8** Complete the energy change for the tube in a TV: Electrical to ______.

**9** Figure 1.14 shows the input signal to an amplifier. Draw the output signal.

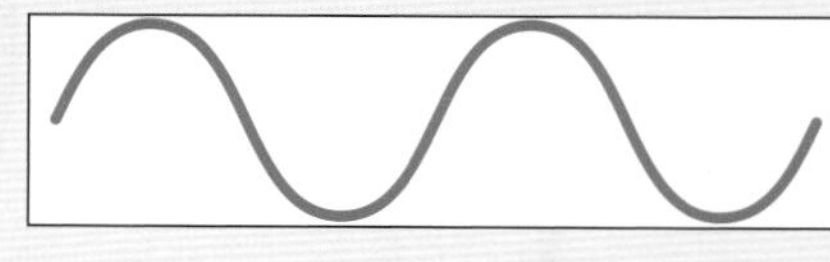

**Figure 1.14**

**10** When you push the button on the remote control to change channels, which part of the TV are you operating?

# 3 Satellites

## After studying this section you should be able to...

1 Describe how satellites are used in communications.
2 State that a geostationary satellite stays above the same point on the Earth's surface.
3 State that curved reflectors on receiving aerials make the signal stronger.
4 Explain why curved reflectors on receiving aerials make the signal stronger.

The science fiction writer Arthur C Clarke suggested the idea of satellites in 1946. He thought that just three satellites could cover the Earth (figure 1.15).

A satellite is any object that goes around a large object. The Moon is a satellite of the Earth but it cannot send any signals.

Satellites are used to send lots of information from one part of the world to another. We need satellites because we cannot just send the signals straight from the UK to the USA. This is because:

- The signals from transmitters travel in straight lines. This happens with high frequency TV signals.
- The Earth is curved and these TV signals will not travel directly from Scotland to the USA.

**Figure 1.15** *Satellites orbiting the Earth.*

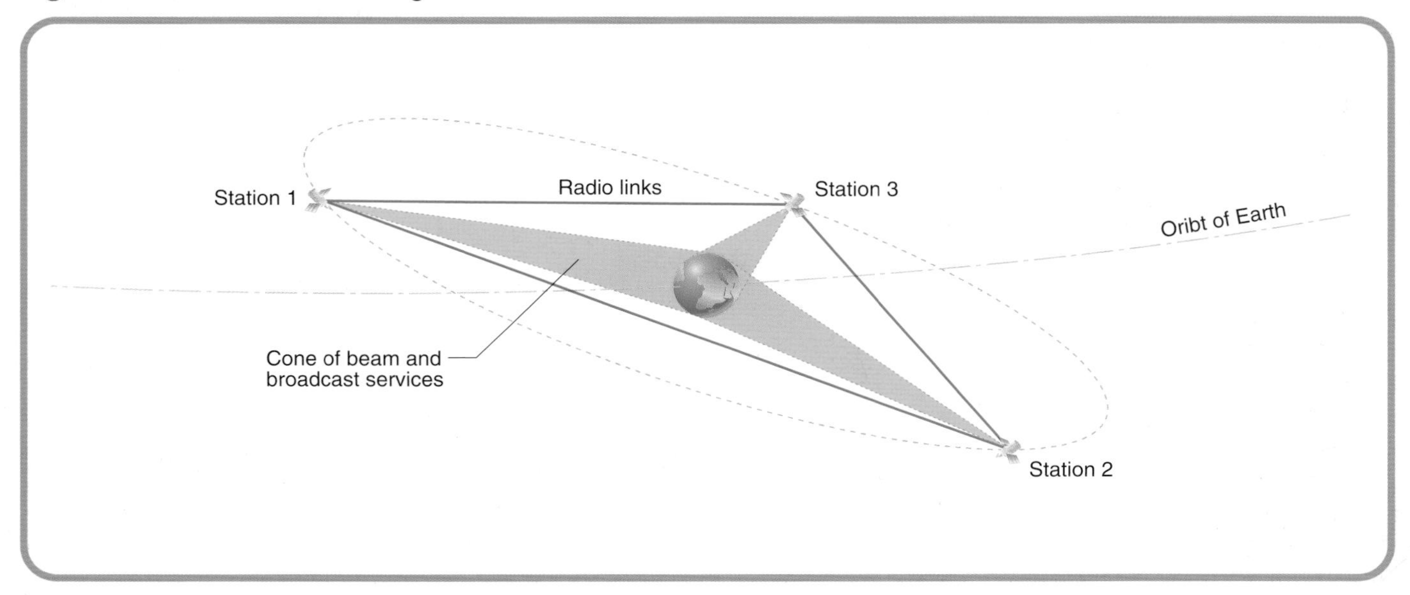

## Communication satellites

The world makes use of satellites to communicate in a number of ways:

- For telephone and TV programmes.
- For weather information to use for forecasts.
- For checking on crops and changes in the Earth.
- For information on the security of countries using spy satellites.

A typical satellite is shown in figure 1.16, and the details of the parts in figure 1.17.

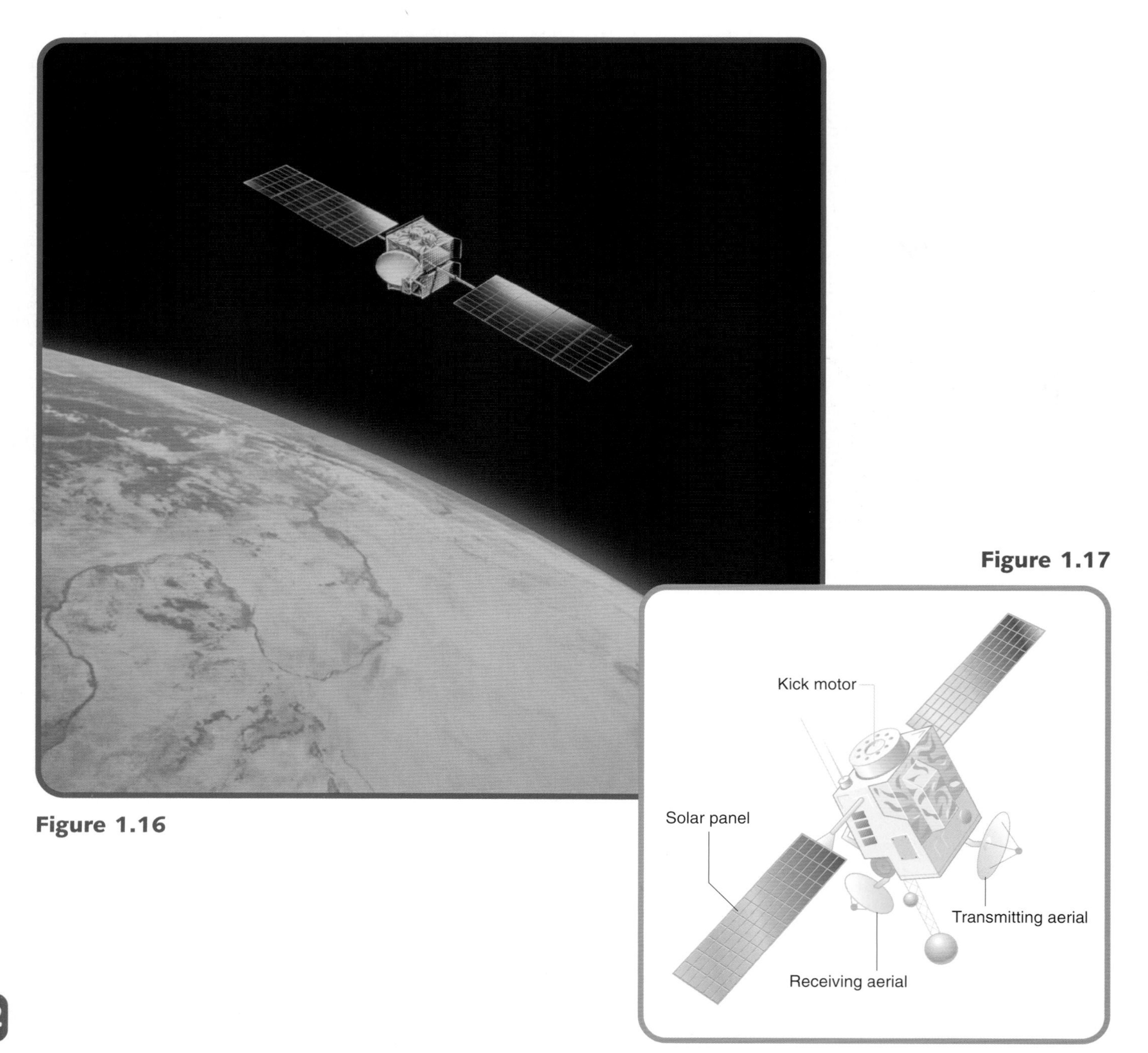


Figure 1.16

Figure 1.17

## Receiving signals from satellites

A satellite has a very sensitive receiver as the signals have to travel about 36 000 kilometres to reach it. The signal when received and focused is amplified and transmitted back to Earth. Very large dish aerials on the Earth are required to pick up the weak signals coming from satellites (figure 1.18). A typical satellite dish is shown in figure 1.19.

**Figure 1.18**

**Figure 1.19**

## Geostationary satellite

The time taken for a satellite to orbit around the Earth is called its period. This depends on the satellite's height above the Earth's surface. The further the satellite is from the Earth, the slower it appears to move. It has been calculated that a satellite in orbit 36 000 kilometres above the equator would complete one orbit in the same time as the Earth revolves, which is 24 hours. This is called a geostationary satellite. It appears to be stationary above the same point on the Earth. Satellites in geostationary orbit must be in orbit above the equator.

Early satellites were not able to be put into a high orbit so they orbited the Earth in less than 24 hours.

In 1965 Early Bird was placed in a geostationary orbit. This was the first geostationary satellite used for communication.

## Using dish aerials (curved reflectors)

When radio broadcasting began in 1920, the medium frequency (MF) radio signals then used could travel about 1600 kilometres. Soon after high frequency (HF) radio bands were discovered these were used for worldwide communication.

Receiving dishes gather in most of the signal and reflect it to one point called the focus. The receiving aerial is placed at the focus to receive the strongest signal (figure 1.20).

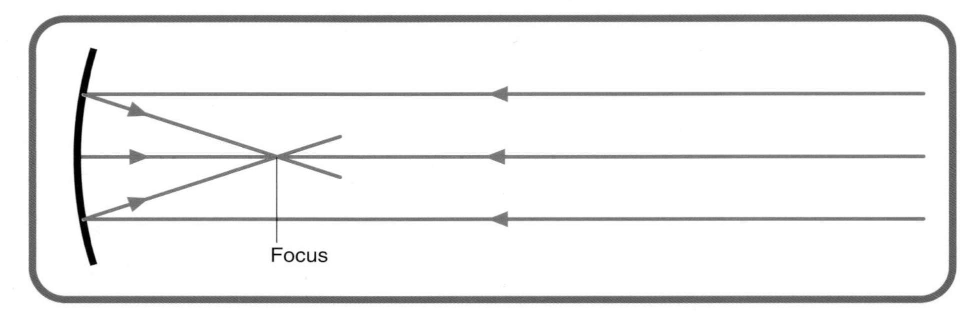


**Figure 1.20**

The signal received is then amplified and transmitted as a narrow parallel beam back to Earth. Panels of solar cells provided the energy to do this.

Satellite communication systems have expanded rapidly. Each new satellite is an improvement on the previous ones. Early Bird could relay 240 telephone calls at once. Modern satellites have several thousand telephone, data and television channels.

The satellite does not act as a mirror, but is a combined receiver and transmitter.

### Key Points

- Satellites are used for television and telephone communications.
- A geostationary satellite stays above the same point on the Earth's surface.
- Curved reflectors on receiving aerials make the signal stronger by bringing all the signal to one point.

## Section Questions

**1** State two uses of satellites in communications.

**2** What kind of satellite stays above the same point on the Earth?

**3** How long does this kind of satellite take to go round the Earth?

**4** The dishes used for the new TV channels are curved. Why are they curved rather than flat?

**5** Copy and complete figure 1.21 to show the signals after they reach the reflector.

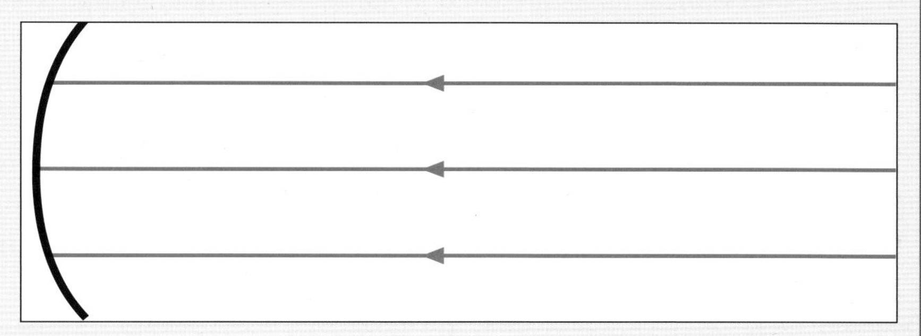

**Figure 1.21**

**6** When the signal from the transmitter reaches the satellite, why does it have to be amplified?

**7** What part in a satellite provides electrical power to operate the satellite?

**8** Why do we need satellites to send signals from Scotland to South America, rather than sending them directly across the world?

**9** Use the graph in figure 1.22 to estimate the height of a weather satellite which takes 12 hours to orbit the Earth.

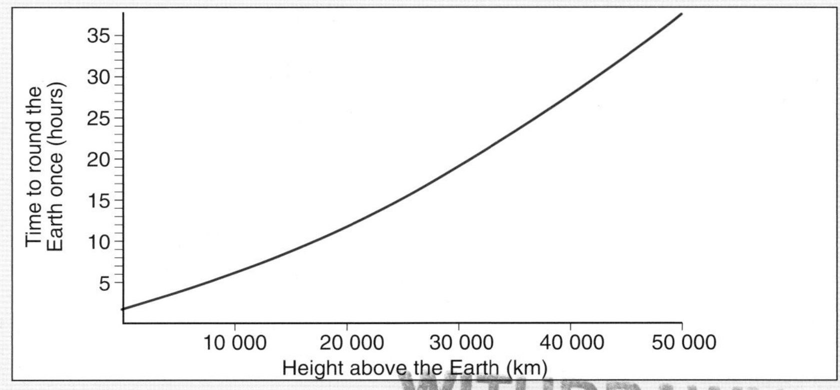


**Figure 1.22**



**10** What is the speed at which signals are sent from Earth to any satellite?

# Optical fibres

## After studying this section you should be able to...

1. State that light can be reflected.
2. Describe the direction of the reflected light from a plane 'mirror'.
3. State what is meant by an optical fibre.
4. State that optical fibres are used in some telecommunication systems.
5. State that optical fibres transmit light signals.
6. State that signal transmission along an optical fibre takes place at a speed of nearly 200 000 000 metres per second.
7. Describe the transmission of the light signal along an optical fibre.
8. State that many telecommunication links into the home are by optical fibres.
9. Describe one advantage and one disadvantage of using optical fibres for transmission of signals into the home.

## Reflection of light

To understand how optical fibres operate we must look at the reflection of light. We see objects because the objects give out light or they reflect some of the light falling on them. Every object will reflect some light but the best reflectors are those that are smooth and highly polished such as mirrors.

## Law of reflection

A ray of light is shone from a ray box at a mirror. The angle of the ray is measured from a line drawn at right angles to the surface called a normal. This angle is the angle of incidence.

The angle of the reflected ray is measured in this way. This is the angle of reflection. This measurement can be repeated for other angles of incidence.

It is found that the angle of incidence is always equal to the angle of reflection (figure 1.23).

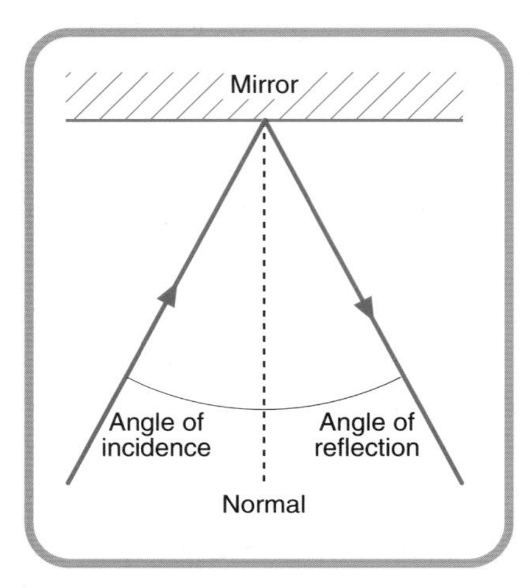


**Figure 1.23**

**Figure 1.24** *Optical fibres.*

## Sending messages with light

### Optical fibres

Optical fibres are made of a thin flexible glass thread which is about the thickness of a human hair. Most of the glass thread is an outer layer of glass called cladding glass. In the middle of the thread is a different glass which is very, very thin. The fibres are made of extremely pure glass to cut down light loss. The fibres have a protective coating which reflects the light — keeping it inside the fibre.

The glass is so pure that a very thick block of it would be as clear as an ordinary window. We could see down to the bottom of the sea (figure 1.24)!

### Why we use optical fibres

- Optical fibres are *lighter*.
- They carry *more information* and give better quality communications than normal telephone wires.
- The signal that passes along the fibre is not electrical, so it is less likely to be affected by other people's telephone calls or by other forms of electrical interference.
- They are cheaper to make than copper, since glass is cheaper than copper.
- The **disadvantage** is that it is more difficult to join fibres together than copper wires.

Many modern telecommunication systems use optical fibres instead of copper wires. One single hair-like optical fibre can carry all the information needed to bring telephone messages, cable TV and computer services into your home.

The fibres operate by light being reflected down the fibre. No light can leave the outside of the fibre (figure 1.25). It has been estimated that one optical fibre cable could take all the telephone calls being made at once in the world.

**Figure 1.25**

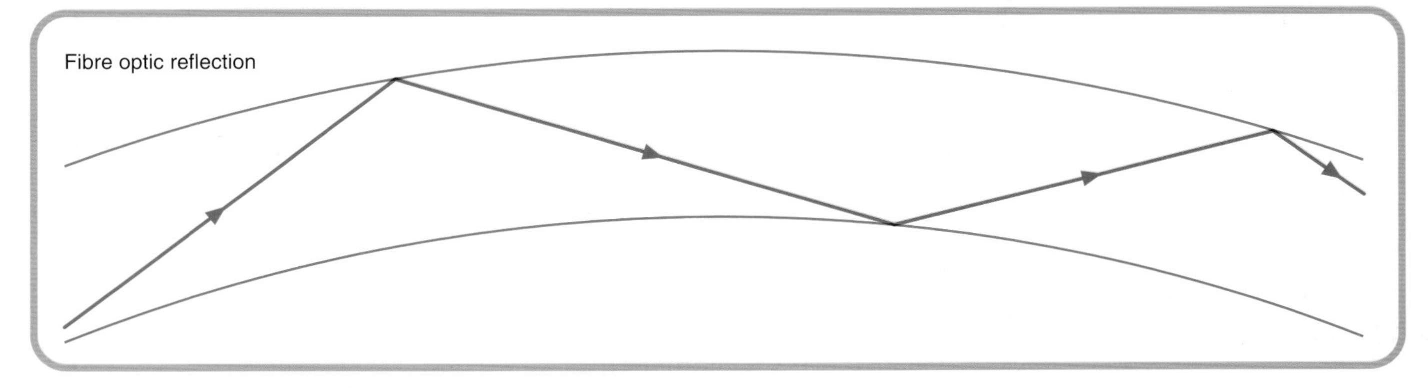

### Transmission and detection

To send speech information along glass fibres it is first necessary to change sound signals into suitable pulses of electrical energy. The microphone (transmitter) in a telephone hand-set and electronics do this. These pulses of electricity control a small laser (a narrow, very powerful beam of light — see chapter 3) or an LED (see chapter 4). These then produce pulses of light which are transmitted through the optical fibre (figure 1.26). The speed of light in an optical fibre is 200 000 000 metres per second.

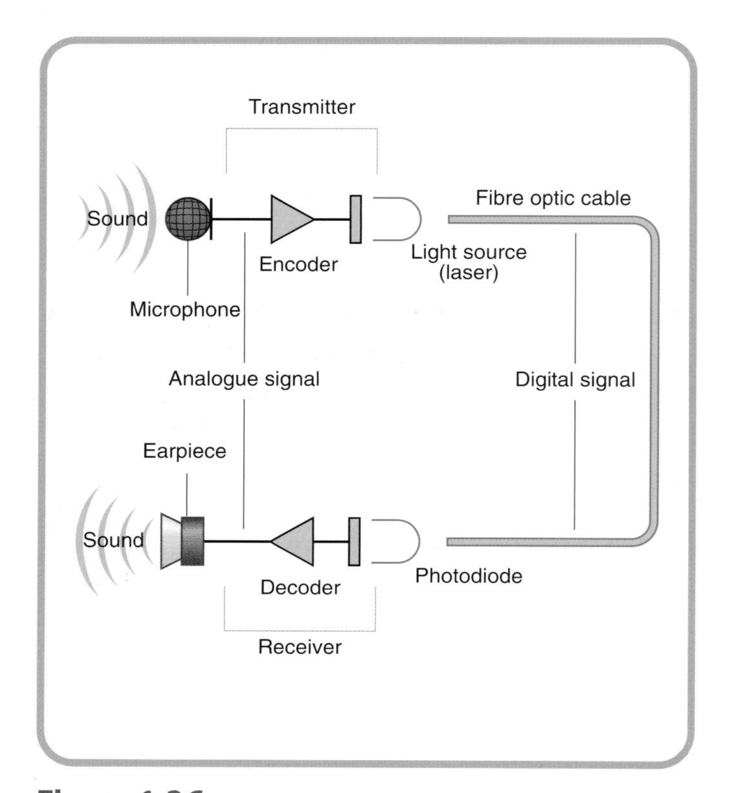


**Figure 1.26**

**Figure 1.27** *Fibre optic cables being laid.*

At the receiving end the light pulses are changed back into pulses of electricity by a small device called a photodiode. The electrical signal is fed to the earpiece of the telephone (receiver) which then reproduces the original sound. All of Britain's major cities are now linked by a major fibre link trunk system. Cable TV operators also use these fibres for both TV and telephone communications. A transatlantic optical fibre link is now also available. Fibre optic cables are laid as shown in figure 1.27.

Modern optical fibres transmit light signals with very little loss of energy from the signal. They can be used over distances of about 100 kilometres without being amplified. With conventional copper cables there is so much loss of energy of the signal that amplifiers have to be installed every 4 kilometres.

## Key Points

- Light can be reflected.
- The direction of the reflected light from a plane 'mirror' is the opposite to the direction of the incident ray.
- An optical fibre is a thin piece of glass.
- Optical fibres are used in some telecommunication systems and transmit light signals.
- Signal transmission along an optical fibre takes place at a speed of nearly 200 000 000 metres per second.
- The transmission of the light signal along an optical fibre works by reflection inside the fibre.
- Many telecommunication links into the home are by optical fibres.
- One advantage of using optical fibres for the transmission of signals into the home is that they are cheaper.
- One disadvantage of using optical fibres for transmission of signals is that they are difficult to join together.

## Section Questions

**1** Figure 1.28(a) shows light reaching a mirror. Copy and complete the diagram to show what happens to the light after reaching the mirror.

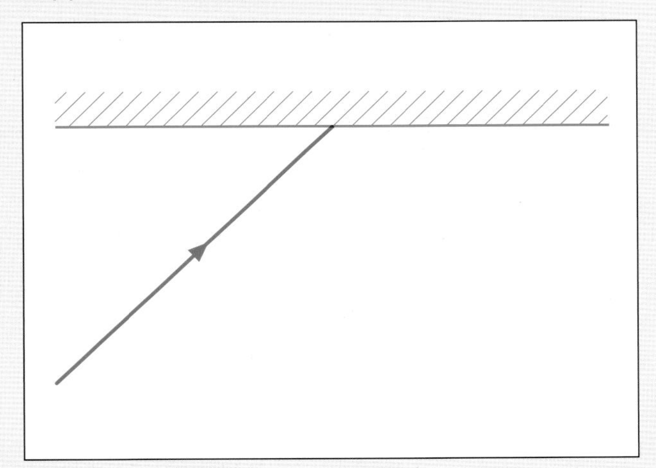

**Figure 1.28(a)**

**2** Copy and complete the sentence 'an optical fibre is a thin ______'.

**3** Optical fibres are used in communication systems. What type of signal is sent down these fibres?

**4** Copy and complete figure 1.28(b) to show how light is sent down the optical fibre.

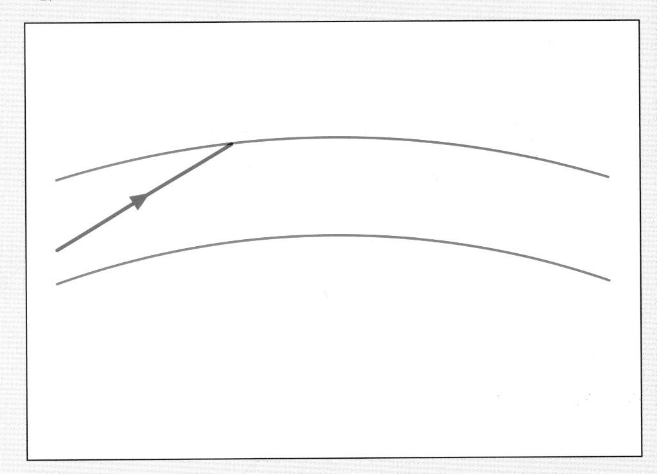

**Figure 1.28(b)**

**5** An advert says that optical fibres are better for communications than copper cables. A friend does not believe the advert. What reasons would you give to try and convince him?

# 5 Telephones

## After studying this section you should be able to...

1. State that in a telephone coded messages or signals are sent out by a transmitter and are picked up by a receiver.
2. State that telephone communication may use metal wires, optical fibre or radio waves between transmitter and receiver.
3. State that a mobile phone acts as a radio transmitter and receiver.
4. State one advantage and one disadvantage of a mobile phone.
5. State that the mouthpiece of a telephone is a transmitter and contains a microphone.
6. State that the earpiece of a telephone is the receiver and contains a loudspeaker.
7. State the energy changes in (a) a microphone (sound to electrical) and (b) a loudspeaker (electrical to sound).
8. State that a telephone signal in a metal wire is transmitted very quickly, at a speed of almost 300 000 000 metres per second.
9. State that fax is the name given to the transmission of documents by telephone communication.
10. State one advantage of using fax.
11. Describe the effect on the signal pattern displayed in an oscilloscope due to a change in (a) loudness of sound and (b) frequency of sound.

## Sending signals

Many communication systems often pass information in code. Native North Americans used smoke signals to communicate news of important events. It was a few years later that Samuel Morse developed his system of dots and dashes to send signals using an electrical telegraph system. This was used for several years. Flashing lamps between ships were used to avoid detection of electrical activity by the enemy. This has now largely disappeared since ships now use satellites to find and check the position of other ships (figure 1.29). Telephone calls are often 'scrambled' to avoid someone listening into them if important information is to be sent between governments. However the use of codes has not disappeared, since many internet sites use a form of code called encryption. This allows sensitive information such as financial details to be put into code when passed to a supplier. This is often denoted by a padlock at the bottom of the internet site. A simple code is the use of text messaging for mobile phones.

**Figure 1.29** *GPS system.*

## The telephone

In 1875, a Scotsman, Alexander Graham Bell, sent the first spoken sentence by telephone.

He was a teacher for the deaf, and had been trying to help people hear messages. When working on his transmitter he spilled some acid on his clothes. He shouted to his friend 'Mr Watson, come here — I want you'. Mr Watson was the first person in the world to hear a message on the telephone. Bell's original machine was a transmitter and receiver. It worked well as a receiver but not as a transmitter.

Any communication system needs a transmitter and a receiver, and the most common is the telephone. This consists of a microphone which acts as a transmitter (figure 1.30). There is also a loudspeaker which acts as a receiver. In addition there will be a buzzer or bell to indicate when a call is received.

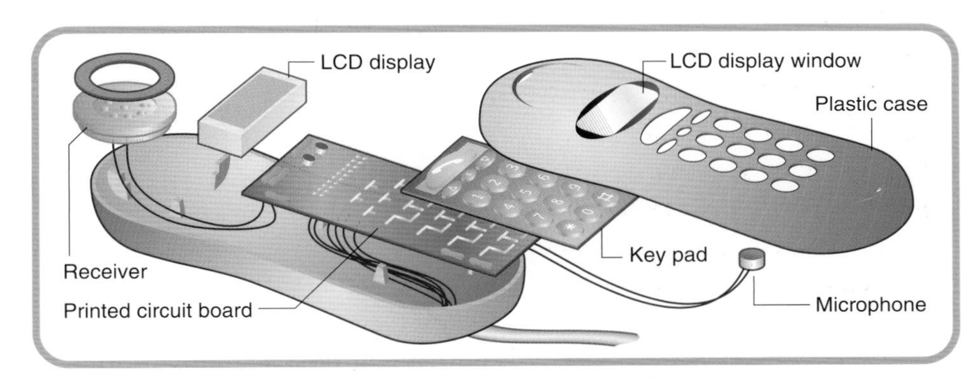


**Figure 1.30**

The key parts of the telephone and the energy changes are shown in table 1.4.

| Part of telephone | Transmitter or receiver | Name of electrical device inside | Energy changes that take place inside |
|---|---|---|---|
| Earpiece | Receiver | Loudspeaker | Electrical to sound |
| Mouthpiece | Transmitter | Microphone | Sound to electrical |

**Table 1.4**

During a telephone conversation electrical signals are sent along the connecting wire at almost the speed of light, that is 300 000 000 metres per second

When you speak into a telephone the sound signal is changed into an electrical signal. These signals can be examined using an oscilloscope. The trace on the oscilloscope shows the electrical patterns of the signals in the wire.

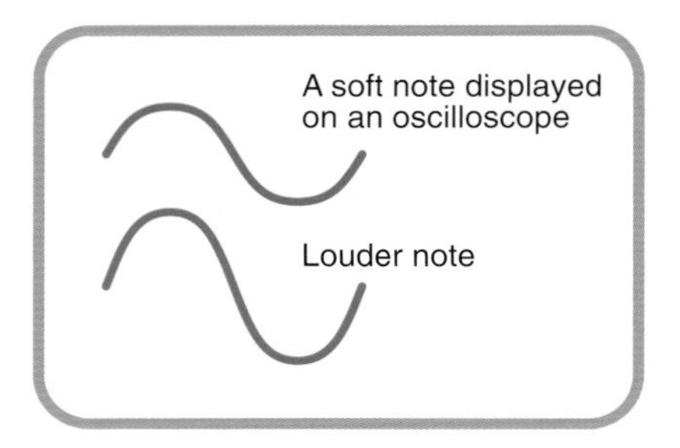


**Figure 1.31**

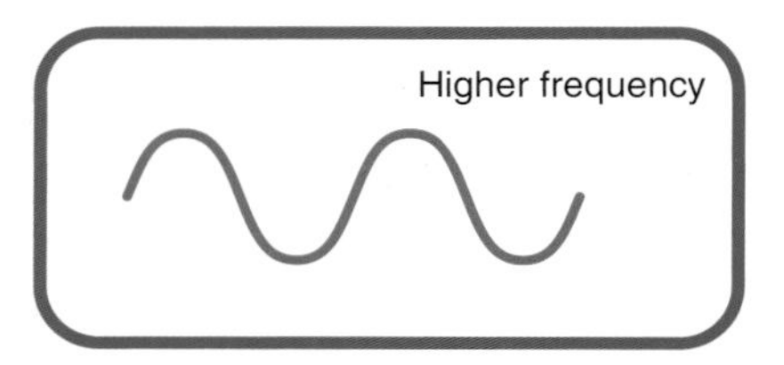


**Figure 1.32**

## Patterns on the oscilloscope

It is possible to display wave patterns for sound signals on an oscilloscope and to see the effect of changes.

If someone sings a note softly into a microphone, the pattern in figure 1.31 can be seen.

If the same note is sung loudly then the sound is increased and the pattern on the oscilloscope changes to the second pattern shown in figure 1.31.

If only the frequency of the note is increased, the pattern changes to that shown in figure 1.32.

These changes show us that:

- When the height of the sound changes, the loudness changes but the frequency is unchanged.
- When the frequency of a sound changes, the note we hear changes but the loudness is unchanged. This means that the number of waves shown on the screen changes.
- The term pitch is used to describe how a noise or musical note sounds to us.
- The higher the pitch, the higher the frequency.

**Figure 1.33**

## Mobile phones

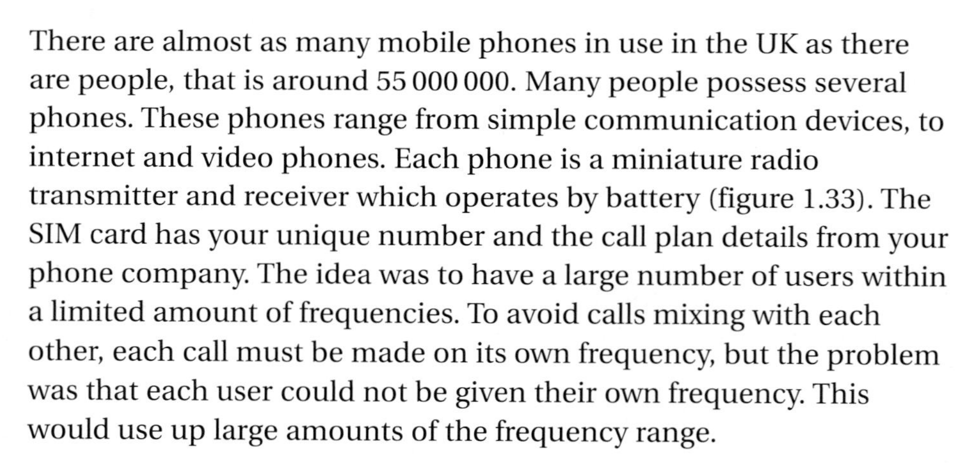

There are almost as many mobile phones in use in the UK as there are people, that is around 55 000 000. Many people possess several phones. These phones range from simple communication devices, to internet and video phones. Each phone is a miniature radio transmitter and receiver which operates by battery (figure 1.33). The SIM card has your unique number and the call plan details from your phone company. The idea was to have a large number of users within a limited amount of frequencies. To avoid calls mixing with each other, each call must be made on its own frequency, but the problem was that each user could not be given their own frequency. This would use up large amounts of the frequency range.

The solution was to divide the country into a number of areas called cells (figure 1.34).

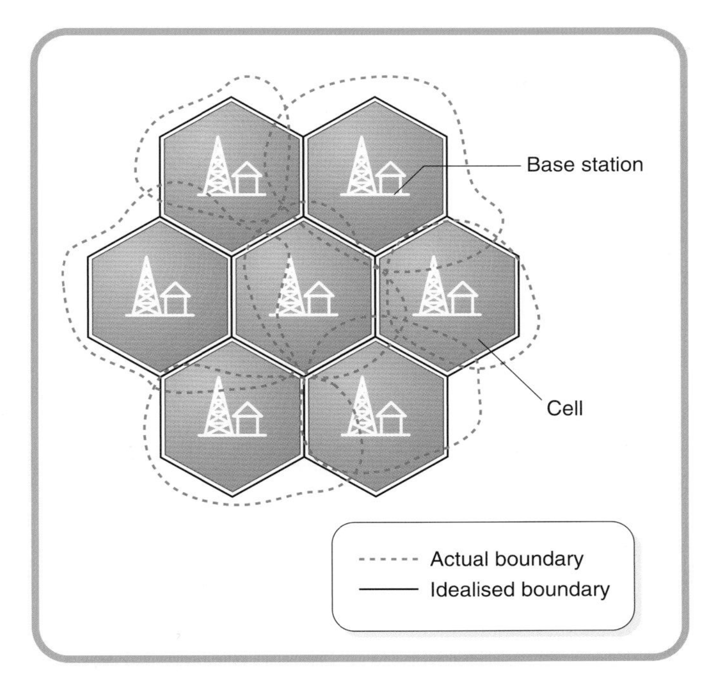


**Figure 1.34** *Division of the country into cells.*

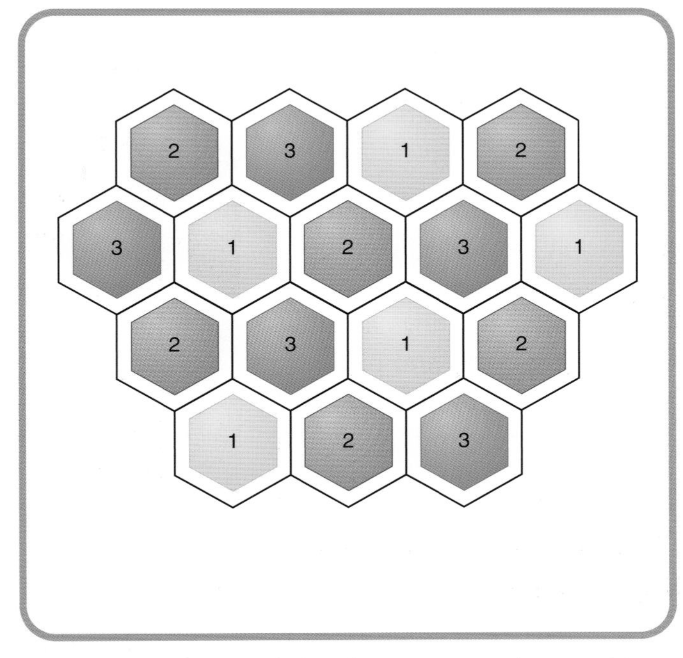


**Figure 1.35** *A mobile phone cannot be used in a neighbouring cell.*

The same set cannot be used in a neighbouring cell, since calls would interfere with each other. They can be used in a cell that is further away (figure 1.35).

In North America, mobile phones are called cell phones.

The frequencies used are very high — above 300 000 000 hertz. These frequencies are used because they have suitable properties:

- A short useful range.
- They work best when there is clear line between the transmitter and receiver.
- The waves can be directed very precisely and need only small aerials.

To help with the increased use of these phones, the cells in a city area are much smaller than those in country areas. The phones change frequencies as you move from one edge of a cell to another, as shown in figure 1.36.

**Figure 1.36** *Frequencies change as you move from one edge of a cell to another.*

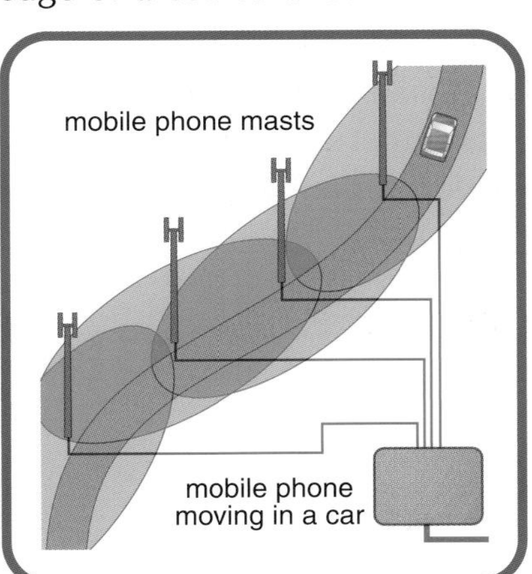


There are some disadvantages of mobile phones:

- They cannot pick up signals in some hilly areas.
- They cannot receive signals inside metal places like a lift or in a tunnel.
- There is concern about health dangers. The microwaves used to carry the signals might cause some brain damage if they are used for long periods of time, since they generate heat.

## Physics for you

### Internet ready and video phones

The latest telephones can transmit not just e-mails but information from a wide range of sites. They can also send video links and pictures (figure 1.37)! A new network has to be built so that large amounts of information can be sent quickly to users.

Figure 1.37

Figure 1.38

## Fax

The term fax is short for facsimile machine. The machine copies exactly whatever is on a piece of paper and then transmits this to another machine through the telephone system. The machine has a special detector which can read changes in light levels as a light passes over the paper. The advantage of this is that complicated drawings can be sent. These are more difficult to send in e-mails.

A machine is shown in figure 1.38. Note that most companies will have a separate number for a fax to their telephone number.

## Key Points

- In a telephone, coded messages or signals are sent out by a transmitter and are picked up by a receiver.
- Telephone communication may use metal wires, optical fibre or radio waves between transmitter and receiver.
- A mobile phone acts as a radio transmitter and receiver.
- An advantage of a mobile phone is the ease of travelling about with the phone. The disadvantage is the signal strength.
- The mouthpiece of a telephone is a transmitter and contains a microphone.
- The earpiece of a telephone is the receiver and contains a loudspeaker.
- The energy change in a microphone is from sound to electrical.
- In a loudspeaker it is from electrical to sound.
- A telephone signal in a metal wire is transmitted very quickly, at a speed of almost 300 000 000 metres per second.
- Fax is the name given to the transmission of documents by telephone communication.
- An advantage of using fax is that images and text can be transmitted.
- The signal pattern displayed in an oscilloscope due to an increase in loudness in sound is the height of the signal increase.
- The signal pattern changes as the frequency of the sound increases and the number of waves on the screen increases.

## Section Questions

**1** Telephones have different designs but they all have two key parts. One part sends out the message and the other part collects it. Name the two parts.

**2** Name one way in which signals are sent between one telephone and another.

**3** For a mobile phone state (a) one advantage and (b) one disadvantage of using this type of phone.

**4** In which two parts of the telephone do these energy changes take place? (a) Electrical to sound, (b) sound to electrical.

**5** At what speed is a telephone signal sent down a metal wire?

**6** John wants to send a copy of a diagram of some wiring to his friend Bill who lives several miles away. Bill needs the diagram very quickly.
a) What device can John use to send the diagram?
b) What is the advantage of using this device?

**7** When a recording is being made, sound engineers replay the recording to check the quality. What difference will they note between a quiet note and a loud note on an oscilloscope screen?

**8** A singer sings a high note and then a low note but at the same loudness. Draw a diagram to show how these two notes would be seen on an oscilloscope screen. The high note is shown in figure 1.39.

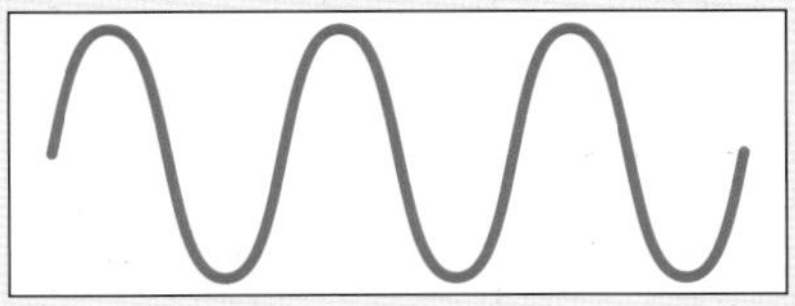

**Figure 1.39**

**9** In mobile phones, the frequency will change as you move from one 'cell' to another. Why is this frequency change needed?

**10** Complete this statement about telephones. Modern telephones use one of three different ways to send signals to another telephone. These are metal ______, optical ______ or radio ______.

## Examination Style Questions

**1** Two friends decide to climb up a hill. They start early one morning and take a small radio to check the weather forecast and a mobile phone. They have to select a station for a local forecast.

a) Which part of the radio will they use?

b) What does the aerial do in the radio?

c) The speed of sound is 340 metres per second and the speed of light is 300 000 000 metres per second. Which of these speeds is the one at which radio waves travel?

d) The weather on the mountain changes quickly. Some people on the mountain phone a friend to let him know that they will be late in reaching him. What is the advantage of the mobile phone?

e) Suddenly the sound from the mobile phone fades. Suggest a possible reason for this.

**2** It is now possible to telephone from some aircraft. This is done using a satellite to receive and transmit the calls.

a) The satellite used is called a geostationary one. What does this tell you about the position of the satellite above the Earth?

b) What energy change takes place in the loudspeaker of the telephone?

c) When the signals reach Earth they are received by a dish aerial as shown in figure 1.40. Use a diagram to explain why an aerial dish of this shape is used.

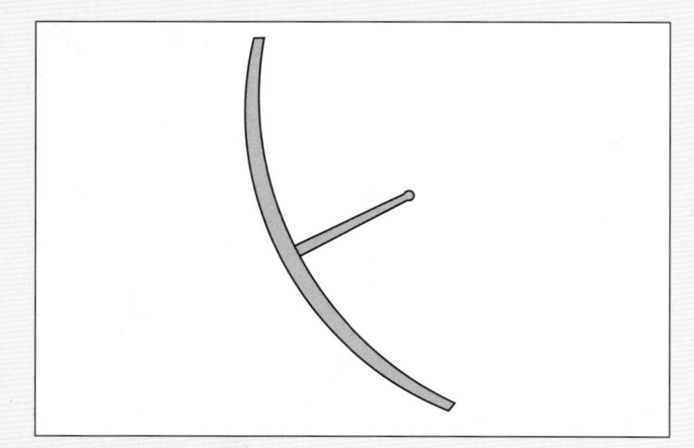

**Figure 1.40**

**3** A block diagram of a television is shown in figure 1.41.

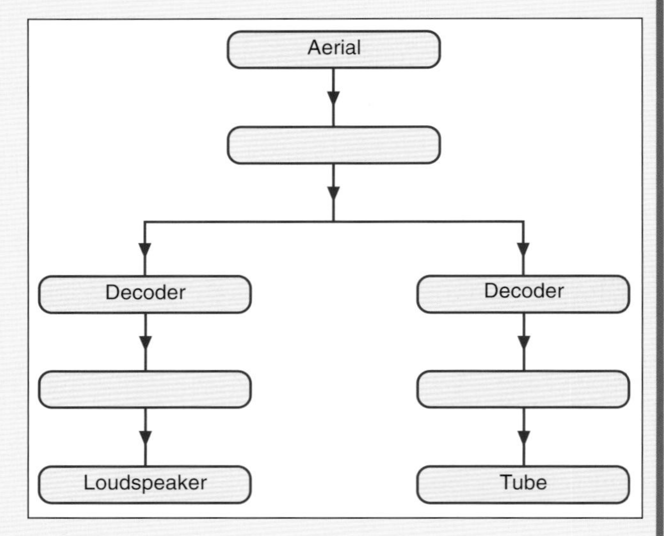


**Figure 1.41**

a) Copy and complete the diagram.

b) What energy change takes place at the tube?

c) Which three colours are used in the screen to make all the colours?

**4** The diagrams used in this book were sent from the artist to the printers.

a) They had to be checked urgently. What device could be used to do this?

b) What advantage does this machine have?

c) The signals are sent down an optical fibre to a telephone exchange. At what speed do the signals travel down the optical fibre?

d) When they reach the exchange they will require amplifying. Figure 1.42 shows how the signal would look before passing through the amplifier. Show what it would like after passing through the amplifier.

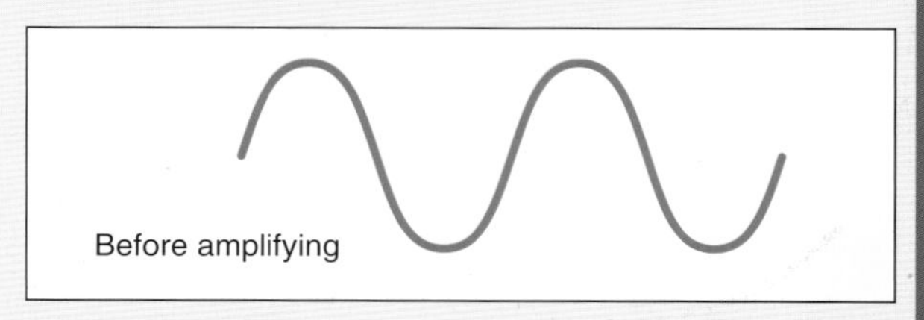


**Figure 1.42**

Unit 2

# Practical Electricity

**We use electricity for many everyday things such as cooking, cleaning, transport and entertainment. The types of electrical circuits used and the measurement of current, voltage, resistance and power are discussed. The safe use of mains electricity is also considered.**

# 1 Electrical circuits

## After studying this section you should be able to...

1. Draw and identify the circuit symbols for a battery, lamp and switch.
2. State that a battery is a source of electrical energy.
3. State that lamps, heaters and motors convert electrical energy into other forms.
4. Describe a series circuit.
5. State that an ammeter is used to measure current.
6. Draw and identify the circuit symbol for an ammeter.
7. Draw a circuit diagram showing the correct position of an ammeter in the circuit.
8. State that current is a flow of charge and is measured in amperes.
9. State that in a series circuit, the current is the same at all points.
10. Describe a parallel circuit.
11. State that the sum of currents in two parallel branches is equal to the current drawn from the supply.
12. State that a voltmeter is used to measure voltage.
13. Draw and identify the circuit symbol for a voltmeter.
14. Draw a circuit diagram showing the correct position of a voltmeter in a circuit.
15. State that voltage is measured in volts.
16. State that the sum of the voltages across components in series is equal to the voltage of the supply.
17. State that the voltage across two components in parallel is the same for each component.

## What is electricity?

Electricity is a form of energy associated with stationary or moving charges in a material.

## Electric current

Consider a lamp and a switch connected by wires to a battery. When the switch is closed, the lamp lights — negative charges called electrons from the negative terminal of the battery move through the wires and lamp to the positive terminal of the battery. This movement, or flow, of negative charges is called an electric current (or current for short). A current is a movement of electrons. As the electrons move through the lamp, some of their electrical energy is changed into heat and light.

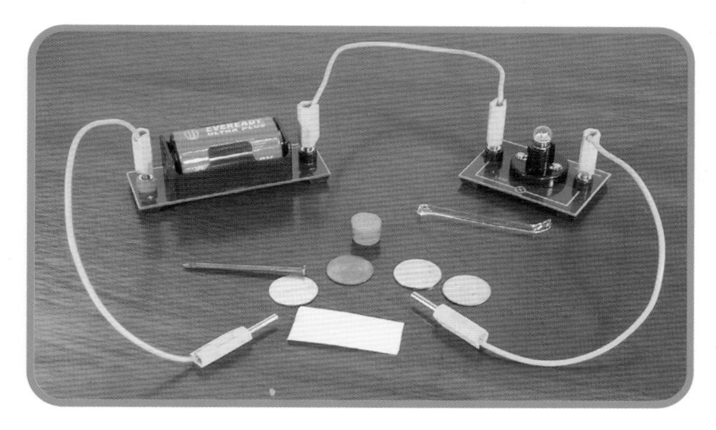


Figure 2.1

## Conductors and insulators

Negative charges (electrons) can only move from the negative terminal to the positive terminal of a battery if there is an electrical path between them. Materials which allow negative charges to move through them easily, to form an electric current, are known as conductors. Materials which do not allow electrons to move through them easily are called insulators.

Set up the circuit shown in figure 2.1. Place different materials in the gap to test whether the material conducts electricity and allows the lamp to light. Make a table of your results using the headings conductors and insulators.

Conductors are mainly metals, such as copper, gold and silver. However, carbon is also a good conductor.

Glass, plastic, wood and air are examples of insulators.

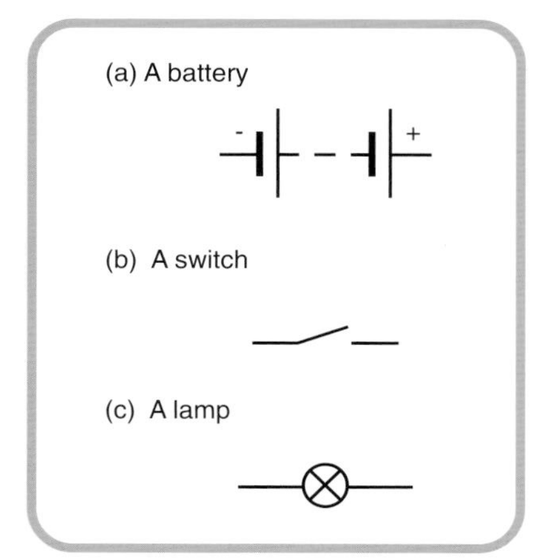


Figure 2.2

## Circuit symbols

So that items or components such as a battery, a switch, a lamp and some wires (conductors) are easily recognised when drawn in a diagram, circuit symbols are used. The circuit symbols for a battery, switch and lamp are shown in figure 2.2. Figure 2.3 shows the circuit diagram for a lamp and a switch connected by wires to a battery.

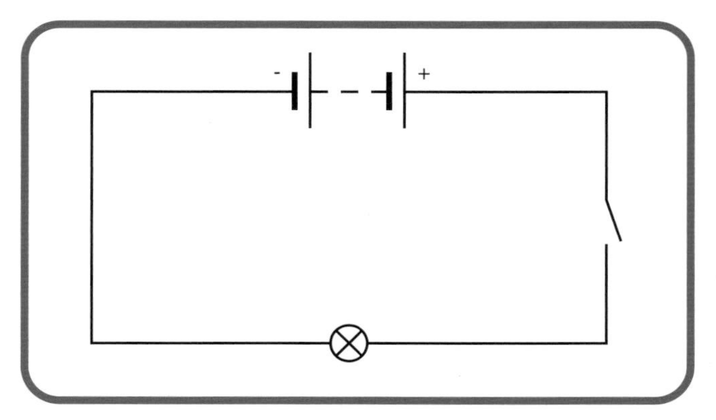


Figure 2.3

The lamp will only light when:

- there is a source of electrical energy — when a battery or power supply is connected.
- there is a complete electrical path (when the switch is closed and all the wires are connected) for the electrons to move round.

## Types of circuit

Electrical components, such as lamps, can be connected in series or in parallel.

Figure 2.4 shows three lamps connected in series. Notice that there is only one electrical path from the negative terminal to the positive terminal of the battery.

Figure 2.5 shows three lamps connected in parallel. Notice that there is more than one electrical path from the negative terminal to the positive terminal of the battery. These alternative paths are called branches.

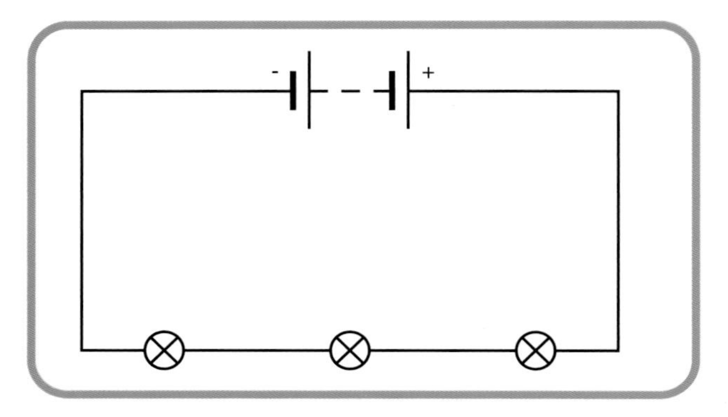


**Figure 2.4**

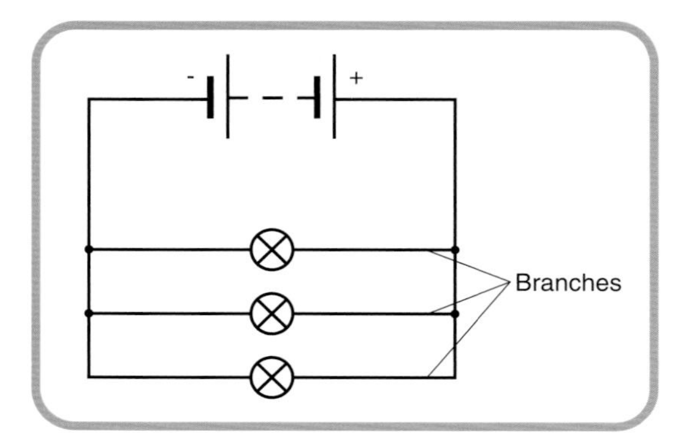


**Figure 2.5**

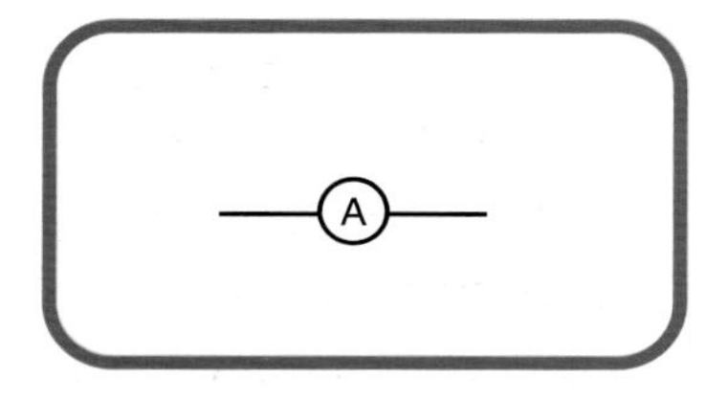


**Figure 2.6**

## Measuring current

An electric current is a flow of charge (negative charges called electrons). Electric current is measured in amperes (A) and an ammeter is used to measure it. The circuit symbol for an ammeter is shown in figure 2.6. Figure 2.7 shows how an ammeter is connected to an electrical circuit.

*Note:* an ammeter measures the current in a component.

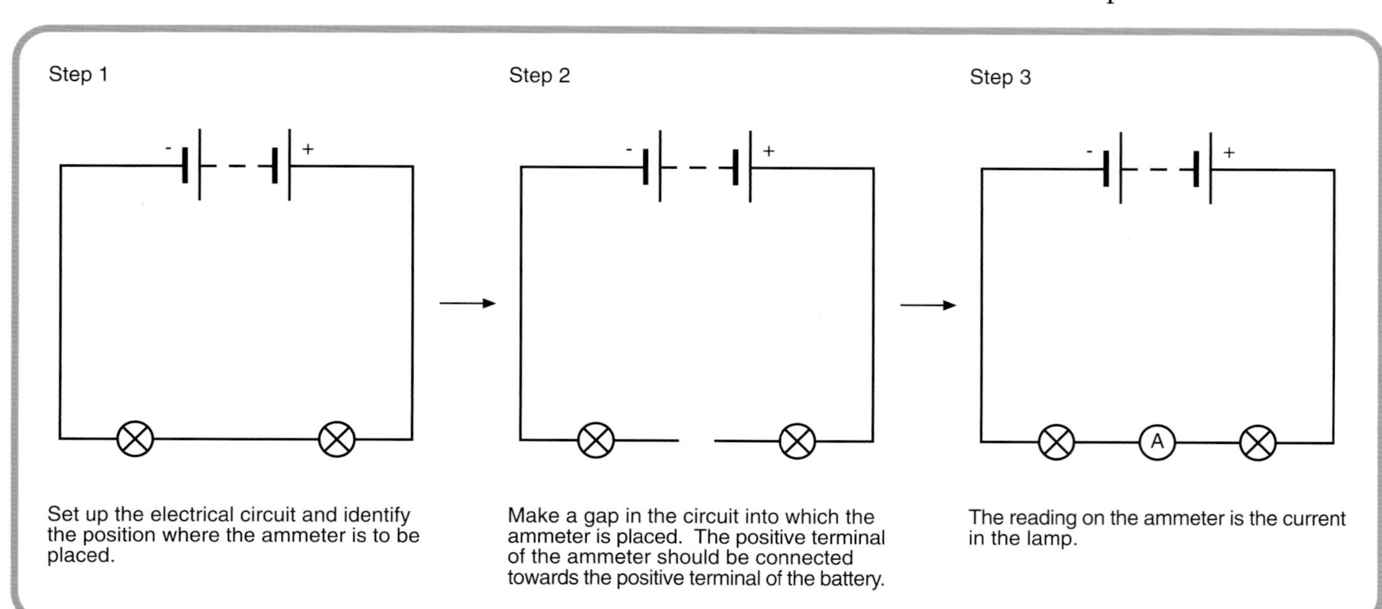


**Figure 2.7**

## Measuring current in a series circuit

Set up a circuit with three lamps connected in series as shown in figure 2.8. Place an ammeter, using the instructions in figure 2.7, to measure the current at position $A_1$. Repeat for positions $A_2$, $A_3$ and $A_4$. What do you notice about the readings on $A_1$, $A_2$, $A_3$ and $A_4$?

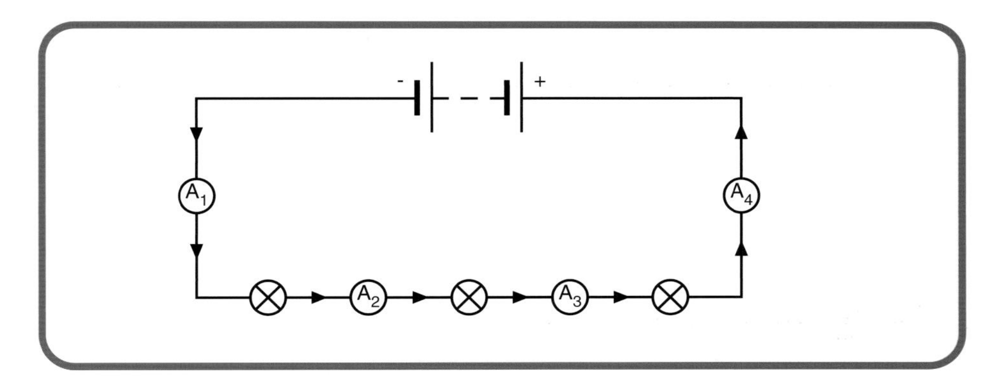


**Figure 2.8**

Figure 2.8 shows three lamps connected in series but with ammeters connected to measure the current at different positions. The readings on the ammeters are recorded as $A_1 = 0.2$ amperes, $A_2 = 0.2$ amperes, $A_3 = 0.2$ amperes and $A_4 = 0.2$ amperes. Notice that the current is the same at each of the different positions. This is true for any series circuit.

For a series circuit the current is the same at all points.

### Examples

Two lamps are connected in an electrical circuit as shown in figure 2.9.

a) Are the lamps connected in series or in parallel?

b) The reading on ammeter $A_1$ is 0.15 amperes. What is the reading on ammeter $A_2$?

Solution

a) Series (as there is only one electrical path).

b) $A_2 = 0.15$ amperes (since lamps are connected in series, then the current is the same at all points).

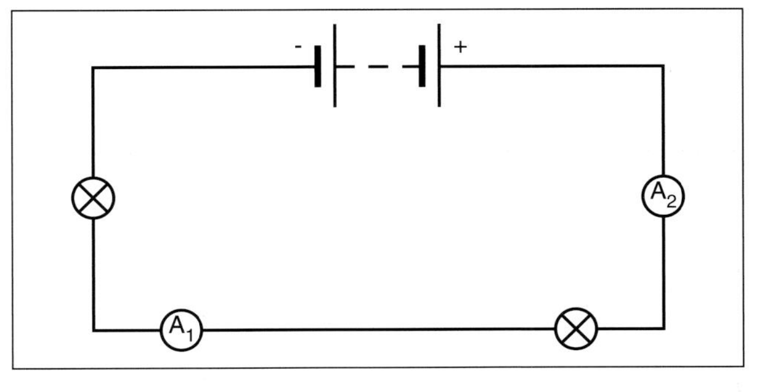


**Figure 2.9**

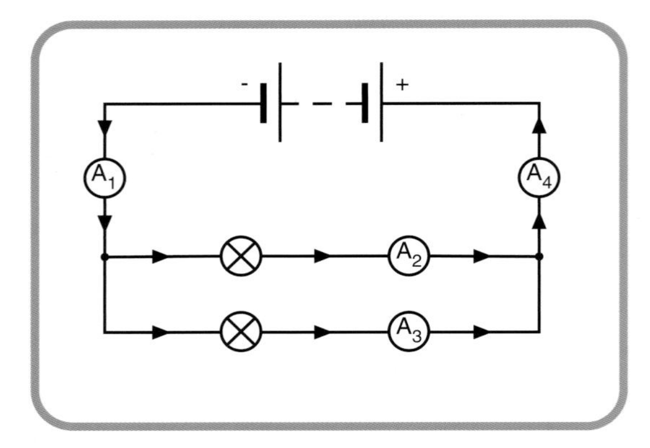

Figure 2.10

## A parallel circuit

Set up a circuit with two lamps connected in parallel as shown in figure 2.10. Place an ammeter, using the instructions in figure 2.7, to measure the current at position $A_1$. Repeat for positions $A_2$, $A_3$ and $A_4$. What do you notice about the readings on $A_1$, $A_2$, $A_3$ and $A_4$?

If you add together the reading on $A_2$ and the reading on $A_3$, what is the answer the same as?

Figure 2.10 shows two lamps connected in parallel with ammeters connected to measure the current at different positions. The readings on the ammeters are recorded as $A_1 = 0.4$ amperes, $A_2 = 0.2$ amperes, $A_3 = 0.2$ amperes and $A_4 = 0.4$ amperes. Notice that the current is smaller in the branches than in the main part of the circuit — the current has split up.

Also $A_1 = A_4$ and $A_1 = A_2 + A_3 = A_4$.

Circuit current = current in the branches added up. This is true for any parallel circuit.

### Examples

Three lamps are connected in parallel as shown in figure 2.11.

What are the readings on ammeters $A_1$ and $A_2$?

Solution
Since the lamps are connected in parallel then:
$A_1 = A_5 = 0.6$ amperes
$A_1 = A_2 + A_3 + A_4$
$0.6 = A_2 + 0.2 + 0.2$
$A_3 = 0.2$ amperes.

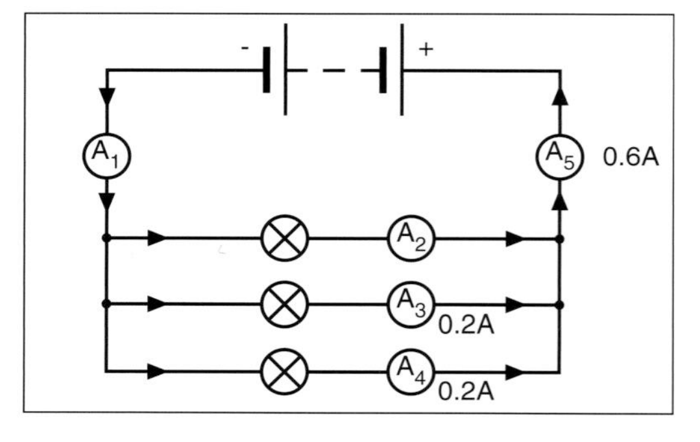

Figure 2.11

## Voltage

Consider a lamp connected to a battery. The battery changes chemical energy, from the substances inside it, into electrical energy. This electrical energy is carried by the charges (electrons) that move round the circuit and is given up as heat and light as they pass through the wire of the lamp. The voltage of the battery is a measure of the electrical energy given to the negative charges passing through the battery.

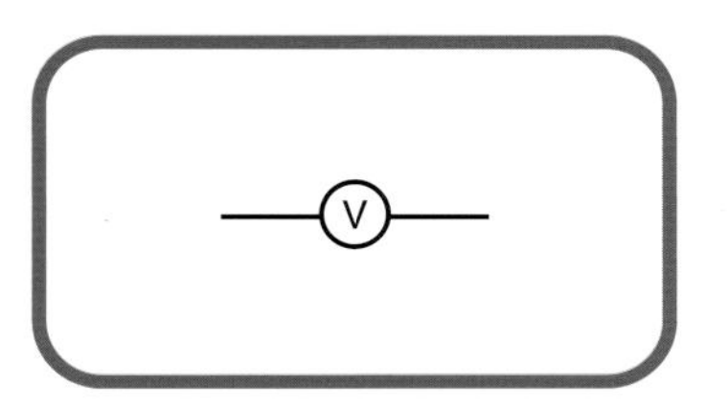


**Figure 2.12**

## Measuring voltage

Voltage is measured in volts (V) and a voltmeter is used to measure it. The circuit symbol for a voltmeter is shown in figure 2.12. Figure 2.13 shows how a voltmeter is connected to an electrical circuit.

*Note:* a voltmeter measures the voltage *across* a component.

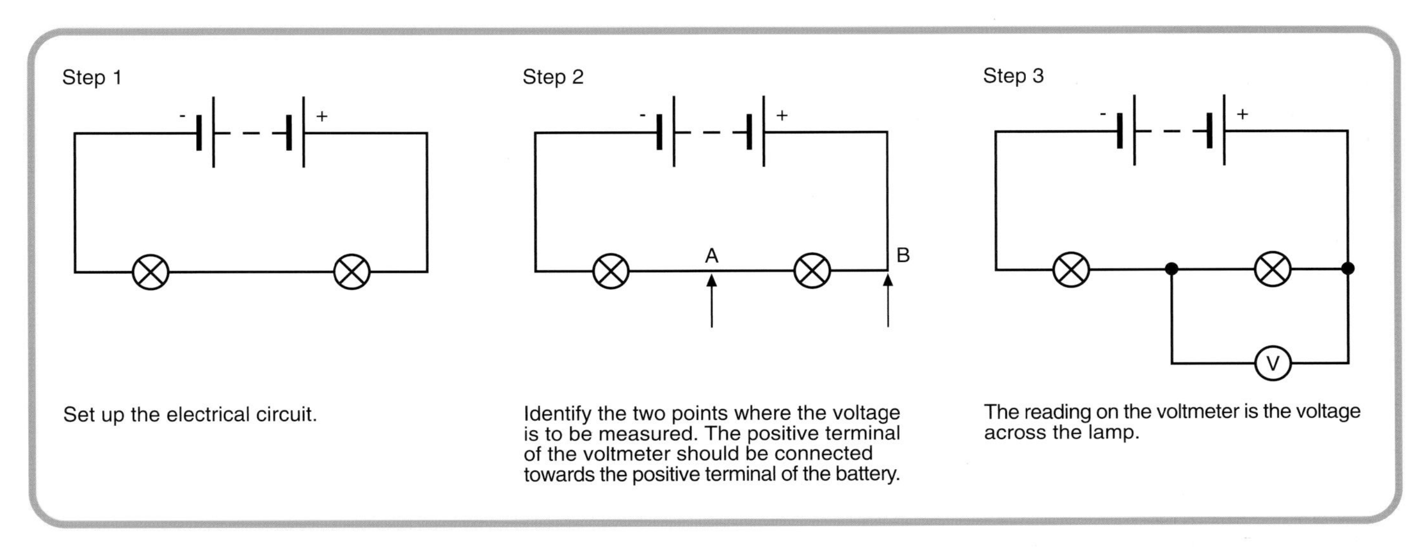


**Figure 2.13**

## Voltage in a series circuit

Set up a circuit with three identical lamps connected in series as shown in figure 2.14. Connect a voltmeter, using the instructions in figure 2.13, to measure the voltage at position $V_1$. Repeat for positions $V_2$, $V_3$ and $V_S$. What do you notice about the readings on $V_1$, $V_2$, $V_3$ and $V_S$?

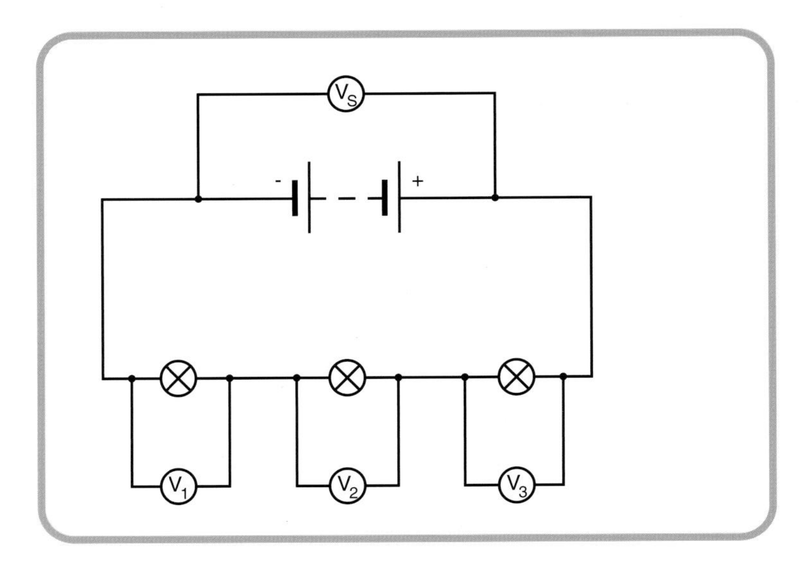


**Figure 2.14**

Figure 2.14 shows three identical lamps connected in series. Voltmeters have also been connected to measure the voltage across each of the lamps and the supply (battery). The readings on the voltmeters are $V_1$ = 1.5 volts, $V_2$ = 1.5 volts, $V_3$ = 1.5 volts and $V_S$ = 4.5 volts.

Notice that the voltages $V_1$, $V_2$ and $V_3$ when added together give the same value as $V_S$. This is true for any series circuit.

The supply voltage is equal to the sum of the voltages round the circuit.

## Examples

Three identical Christmas tree lamps are connected in series to a 9 volt battery.

What is the voltage across each lamp?

Solution

Since the lamps are connected in series and are identical, then the supply voltage is split up equally between the three lamps:

$V_S = V_1 + V_2 + V_3$

$V_S = 3 \times V$

$9 = 3V$

$V_{lamp} = \frac{9}{3} = 3$ volts.

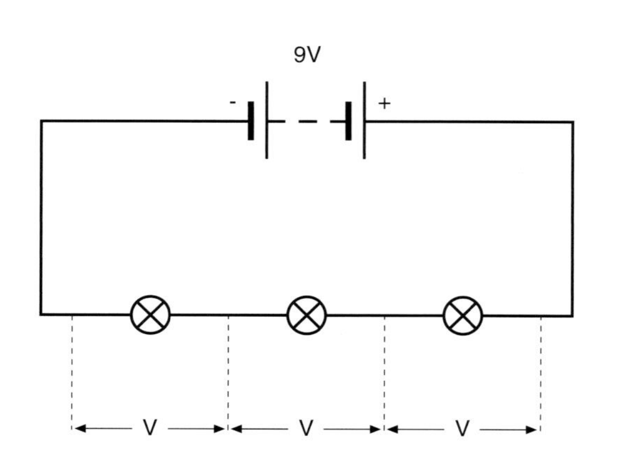


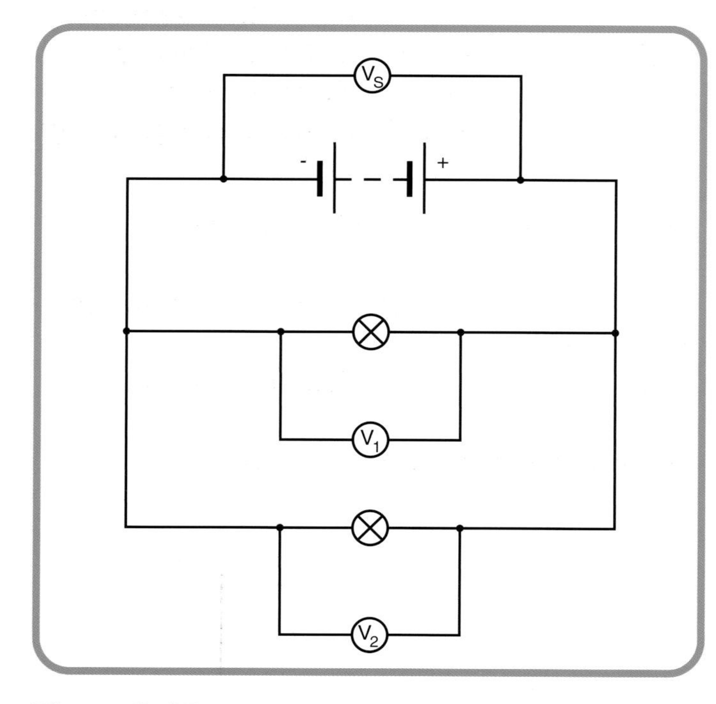


Figure 2.15

## Voltage in a parallel circuit

Set up a circuit with two lamps connected in parallel as shown in figure 2.15. Connect a voltmeter, using the instructions in figure 2.13. to measure the voltage at position $V_1$. Repeat for positions $V_2$ and $V_S$. What do you notice about the readings on $V_1$, $V_2$ and $V_S$?

Figure 2.15 shows two lamps connected in parallel with voltmeters connected to measure the voltage across each lamp. The readings on the voltmeters are $V_1$ = 6 volts, $V_2$ = 6 volts and $V_S$ = 6 volts. The voltage across the lamps connected in parallel is the same, that is $V_1 = V_2$.

Notice that the voltage across parallel branches are the same. This is true for any parallel circuit.

## Examples

Three lamps, P, Q and R are connected to a 4.5 volt supply as shown in figure 2.16. The voltage across lamp P is 1.5 volts.

a) Are lamps P and Q connected in series or parallel?
b) Is lamp R connected in series or parallel?
c) What is the voltage across lamp Q?
d) What is the voltage across lamp R?

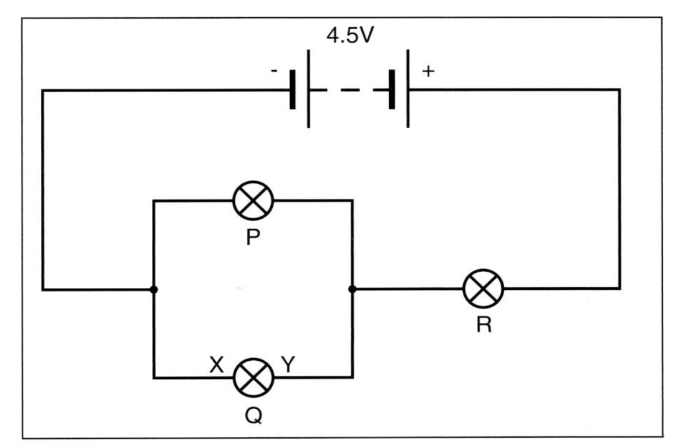

**Figure 2.16**

Solution

a) Parallel, since there is more than one path between X and Y.
b) Series, since only one path in this part of the circuit.
c) P and Q are connected in parallel, therefore they have the same voltage across them and so the voltage across Q = 1.5 volts.
d) The voltage across R is 3 volts since voltages in a series circuit add up.

### Energy conversions

The electrical appliances used in our homes, industry and schools use electrical energy from the mains supply or from batteries. The appliances change electrical energy into a form which is useful for what we are doing and in many cases into other forms of energy which are not so useful. For instance, a lamp changes electrical energy into heat and light. The most useful, and main, energy change for a lamp is the change of electrical energy into light.

The list below shows the main energy change for a number of household appliances:

- The heater element in a kettle changes electrical energy into heat.
- A lamp changes electrical energy into light.
- The motor in a washing machine changes electrical energy into kinetic (movement) energy.
- A radio changes electrical energy into sound energy.

## Key Points

- Current is a flow of charge. Current is measured in amperes (A) using an ammeter. An ammeter is connected in series with an electrical component, such as a lamp.
- Voltage is measured in volts (V) using a voltmeter. A voltmeter is connected in parallel with an electrical component, such as a lamp.
- In the series circuit shown in figure 2.17:
  a) The current is the same at all positions — the current does not split up, so $I_1 = I_2 = I_3 = I_4$.
  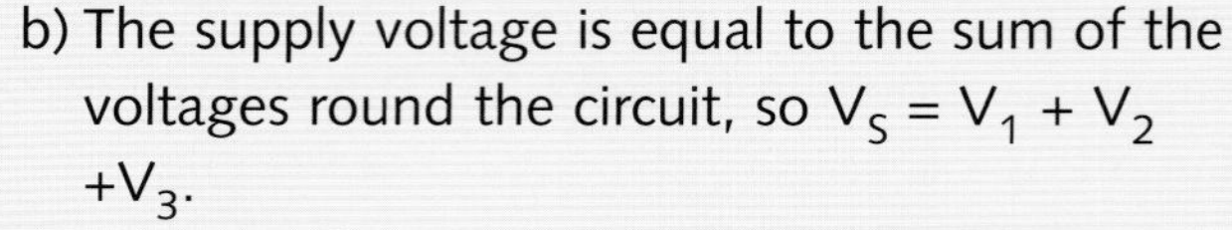
  b) The supply voltage is equal to the sum of the voltages round the circuit, so $V_S = V_1 + V_2 + V_3$.
- For the parallel circuit shown in figure 2.18:
  a) The current splits up so the circuit current equals the sum of the currents in the branches, and $I = I_1 + I_2$.
  b) The voltage across lamps connected in parallel is the same, so $V_1 = V_2$ and in this case $V_S = V_1 = V_2$.

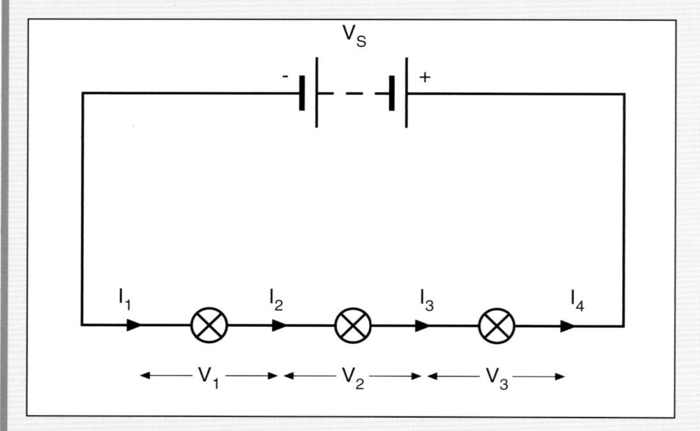

**Figure 2.17**

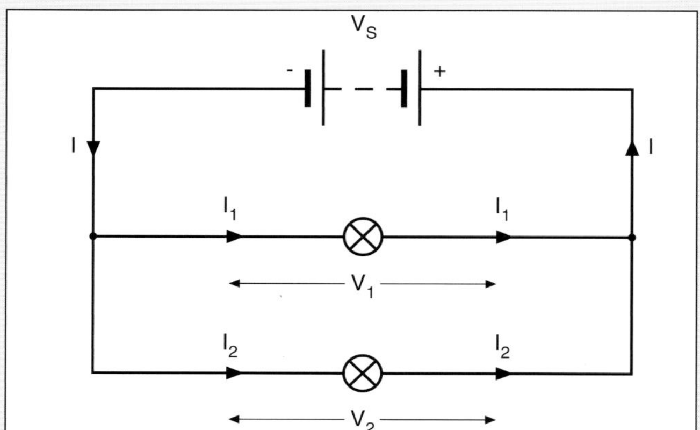

**Figure 2.18**

## Section Questions

**1** Name the components shown in figure 2.19.

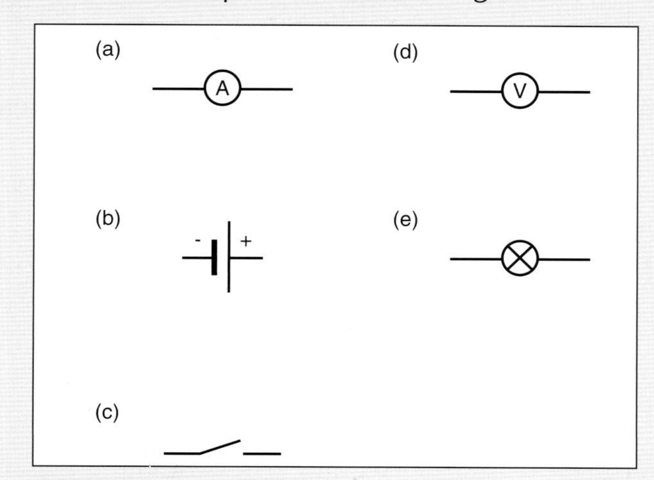

**Figure 2.19**

**2** Copy and complete the following sentences using the appropriate words from the list: amperes, current, electrons, heat, positive, light, source.

A battery is connected to a lamp. The battery is a ______ of electrical energy. Negative charges ( ______) move from the negative terminal to the ______ terminal of the battery. This movement of charge is called a ______. Current is measured in ______. The electrical energy of the battery is converted into ______ and ______ energies by the lamp.

**3** Copy and complete the following sentences using the appropriate words from the list: branches, current, one, parallel, series, voltmeter.

When electrical components are connected in series there is ______ electrical path. In a ______ circuit there is more than one electrical path – the alternative paths are called ______. An ammeter is used to measure ______ and is connected in ______. A ______ is used to measure voltage and is connected in parallel.

**4** For each circuit shown in figure 2.20 state whether the lamps are connected in series or parallel.

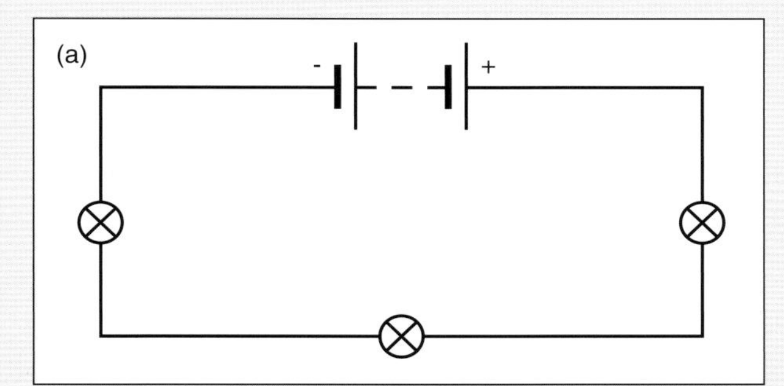


**Figure 2.20a**

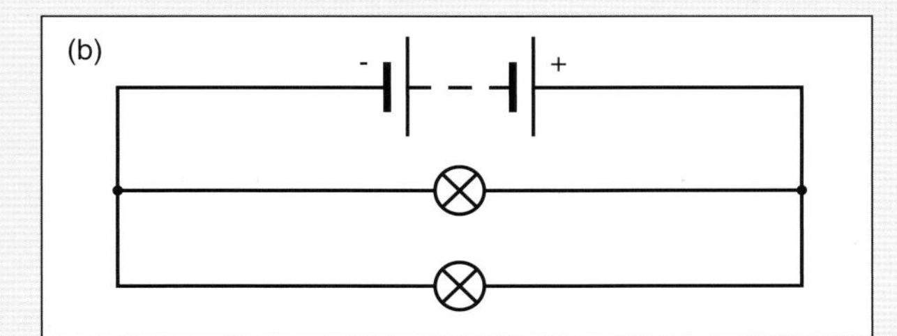


**Figure 2.20b**

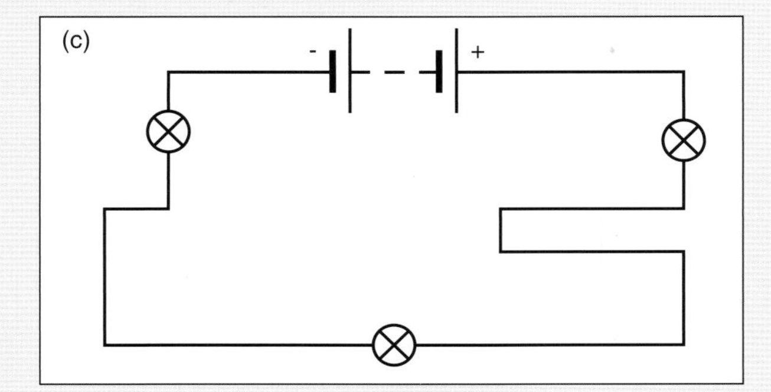


**Figure 2.20c**

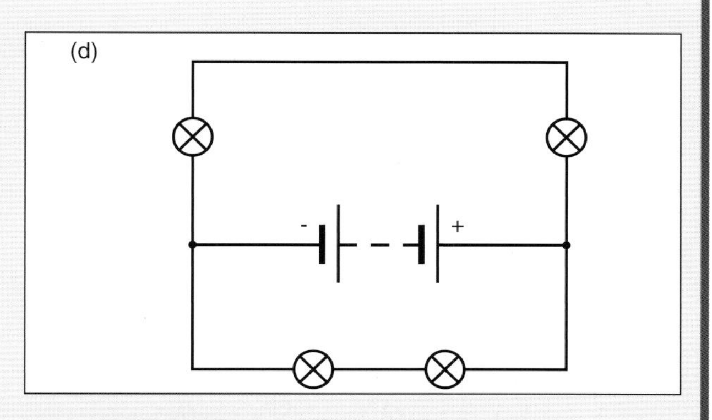


**Figure 2.20d**

**5** A student draws the following circuit shown in figure 2.21. Name the type of meter labelled (a) P, (b) Q.

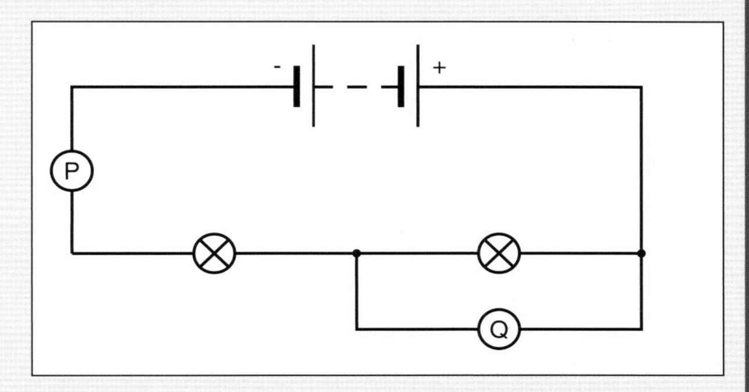


**Figure 2.21**

**6** Redraw each of the diagrams given in figure 2.22 to show how an ammeter is connected to measure the current through lamp Y and a voltmeter is connected to measure the voltage across lamp Z.

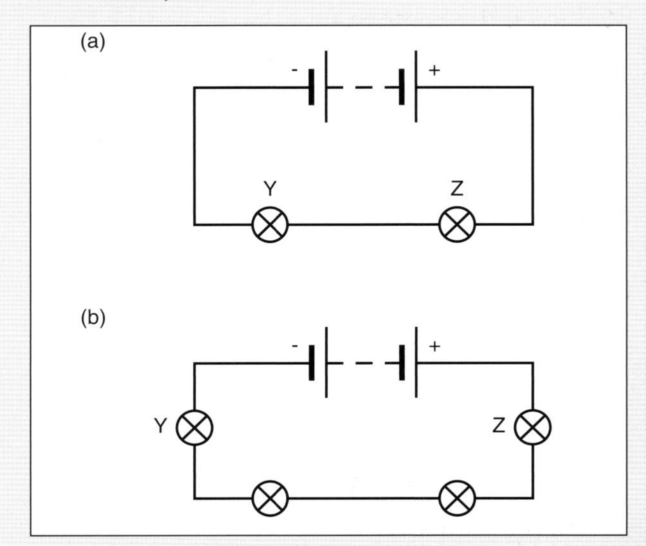


**Figure 2.22**

**7** In the circuits shown in figure 2.23, what are the readings displayed on ammeters:
a) $A_1$, $A_2$, $A_3$,
b) $A_4$, $A_5$ and $A_6$?

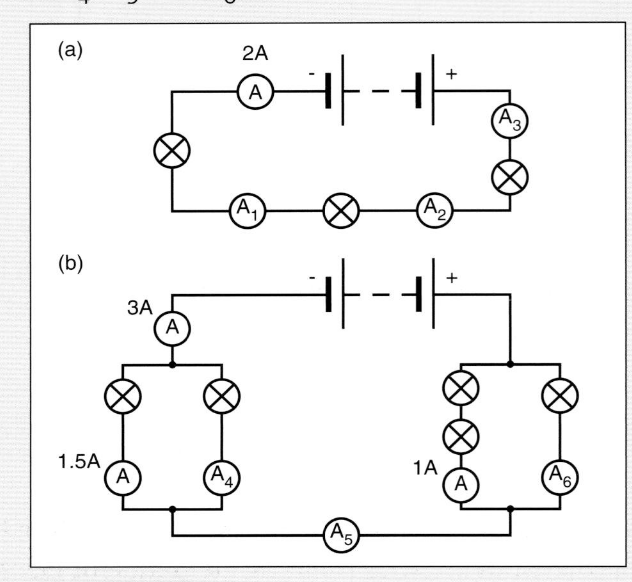


**Figure 2.23**

**8** In the circuit shown in figure 2.24, what is the reading displayed on voltmeter: $V_1$?:

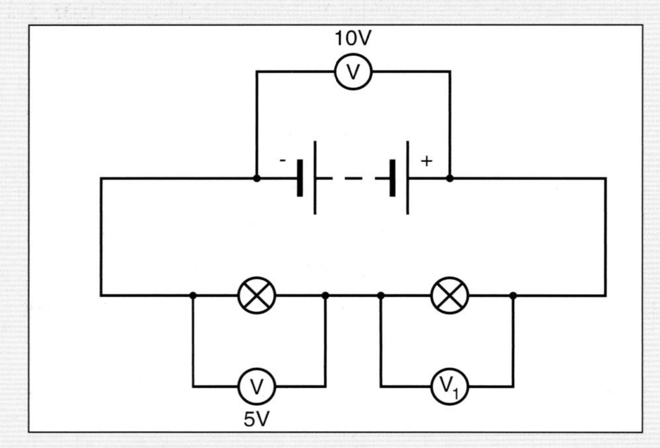


**Figure 2.24**

**9** In the circuits shown in figure 2.25, what are the readings displayed on:
a) ammeters $A_1$ and $A_2$,
b) voltmeters $V_1$ and $V_2$?

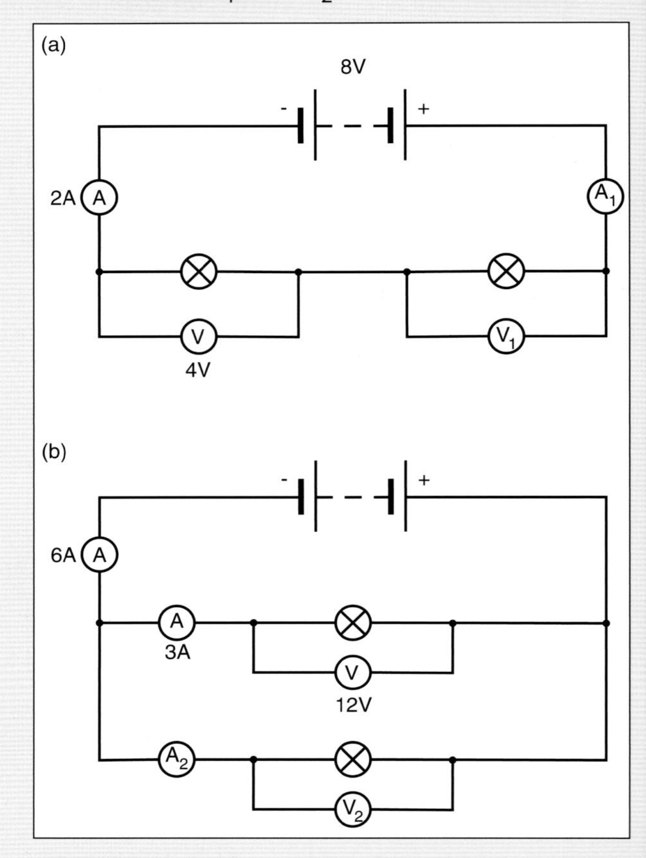


**Figure 2.25**

**10** Draw the circuit symbol for (a) a switch, (b) a lamp, (c) an ammeter, (d) a battery, (e) a voltmeter.

# Resistance

## After studying this section you should be able to...

1. Draw and identify the circuit symbol for a resistor and a variable resistor.
2. State that resistors convert electrical energy into heat energy.
3. State that an ohmmeter is used to measure resistance.
4. State that resistance is measured in ohms.
5. State that an increase in resistance of a circuit leads to a decrease in the current in the circuit.
6. Calculate resistance using resistance = $\frac{\text{voltage}}{\text{current}}$.
7. Give two practical uses of a variable resistor.

When an electric current passes through a wire some of the electrical energy is changed into heat in the wire. Heating elements for electric fires and kettles change electrical energy into heat in the wire inside the element.

All materials oppose current passing through them. This opposition to the current is called resistance. Resistance is measured in ohms (Ω).

The larger the resistance the smaller the current.

The smaller the resistance the larger the current.

For most materials resistance depends on the:

- type of material — the better the conductor the lower the resistance for example copper is a better conductor than iron,
- length of the material — the longer the material the higher the resistance,
- thickness of the material — the thinner the material the higher the resistance,
- temperature of the material — the higher the temperature the higher the resistance.

For a resistor, the resistance value remains constant for different currents, provided the temperature of the resistor does not change.

A resistor whose resistance can be changed is known as a variable resistor. The resistance is normally changed by altering the length of the wire in the resistor (the longer the wire, the higher the resistance). Variable resistors are often used as volume or brightness controls on televisions. Another use for a variable resistor is a dimmer switch for lights — the brightness of the lights is varied by changing the variable resistor control.

An ohmmeter can be used to measure resistance. The circuit symbols for a resistor, variable resistor and an ohmmeter are shown in figure 2.26. Figure 2.27 shows an ohmmeter being used to measure the resistance of a resistor.

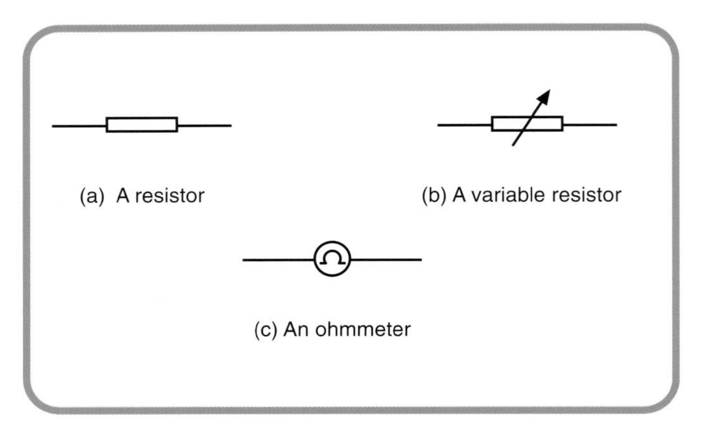


**Figure 2.26**

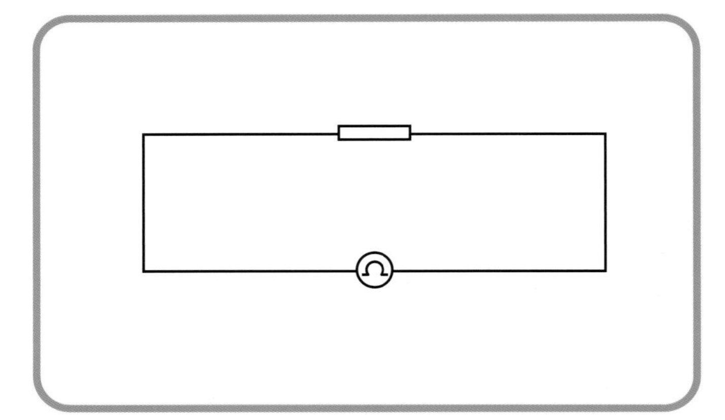

**Figure 2.27**

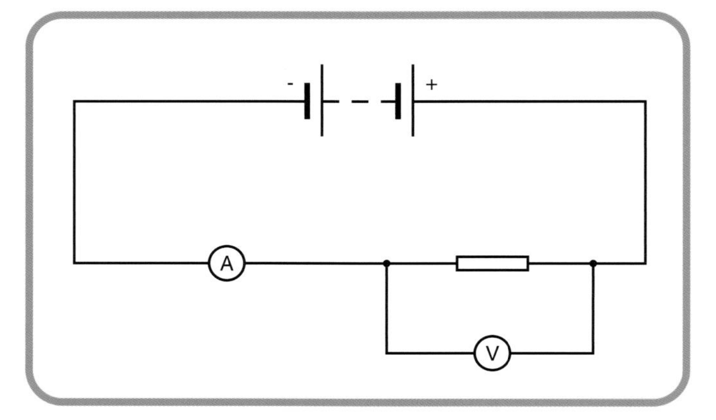


**Figure 2.28**

The resistance of a resistor can also be measured using an ammeter and a voltmeter as shown in figure 2.28. In this method the voltage across the resistor and the current in the resistor have to be measured.

The resistance of the resistor can then be calculated using:

$$\text{resistance of resistor} = \frac{\text{voltage across resistor}}{\text{current through resistor}}$$

that is $R = \frac{V}{I}$.

This is known as Ohm's law.

## Examples

The voltage across a lamp is 12 volts. The current in the lamp is 1.5 amperes. Calculate the resistance of the lamp.

Solution

$R = \frac{V}{I} = \frac{12}{1.5} = 8.0$ ohms.

## Examples

A student sets up the circuit shown in figure 2.29.

a) What is the current in resistor R?

b) What is the voltage across resistor R?

c) Calculate the resistance of resistor R.

Solution

a) 0.25 amperes.

b) 4 volts.

c) $R = \frac{V}{I} = \frac{4}{0.25} = 16$ ohms.

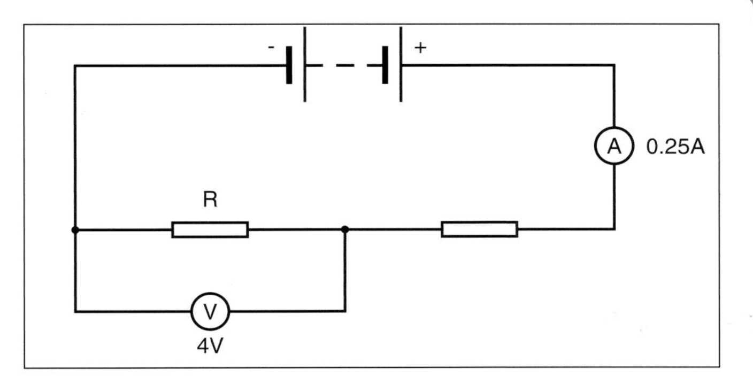


**Figure 2.29**

## Key Points

- Voltage is measured in volts (V), current in amperes (A) and resistance in ohms (Ω).
- The resistance of a resistor can be measured using an ohmmeter.
- Resistance of resistor = $\frac{\text{Voltage across a resistor}}{\text{Current through a resistor}}$

  $R = \frac{V}{I}$.
- In a resistor, electrical energy is changed into heat.
- A variable resistor is a resistor whose resistance can be changed.
- Variable resistors are used as volume controls on radios and televisions. As a dimmer switch the variable resistor is used to alter the brightness of a light.

## Section Questions

**1** Name the components shown in figure 2.30.

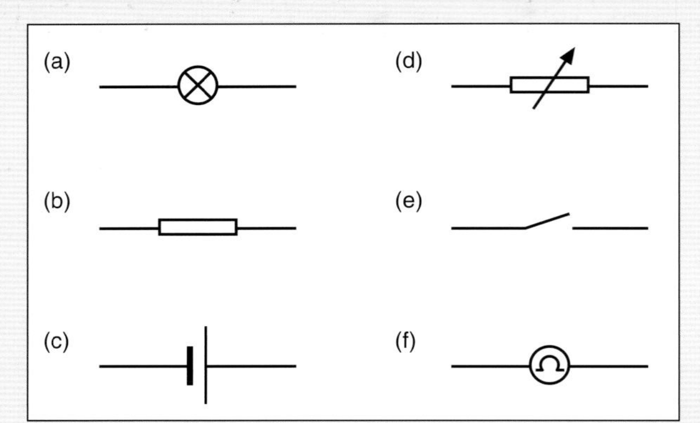


**Figure 2.30**

**2** Copy and complete the following sentences using the appropriate words from the list: decreases, electrical, heat, ohmmeter, ohms, resistance.

The opposition to current is called ______. Resistance can be measured using an ______ and is measured in ______. Increasing the resistance of a circuit ______ the current in the circuit. When a current passes through a resistor ______ energy is changed into ______.

**3** Copy and complete table 2.1.

| | Resistance in ohms | Voltage in volts | Current in amperes |
|---|---|---|---|
| a) | | 10 | 2 |
| b) | | 12 | 3 |
| c) | | 230 | 10 |

**Table 2.1**

**4** A lamp is connected to a 12 volt supply. When lit, the current in the lamp is 4 amperes. Calculate the resistance of the lamp.

**5** An electrical circuit is shown in figure 2.31. Calculate the resistance of the resistor.

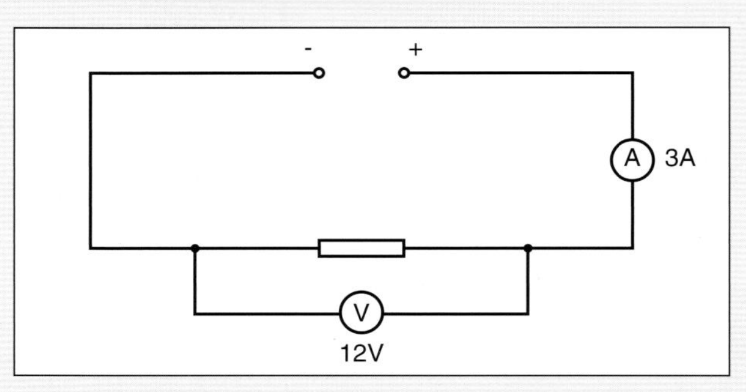


**Figure 2.31**

**6** Two electrical circuits, circuit A and circuit B are shown in figure 2.32.

a) Which circuit shows the electrical components connected in series?

b) Calculate the resistance of resistor X.

c) Calculate the resistance of resistor Y.

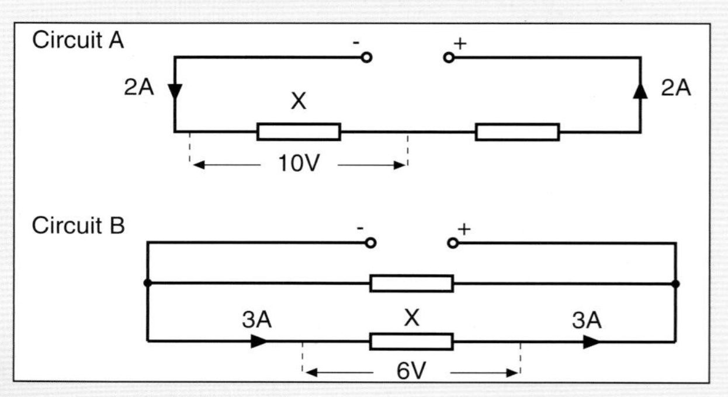


**Figure 2.32**

**7** The element of a cooker is connected to the 230 volt mains supply. The current in the element is 4 amperes.

a) What is the resistance of the cooker element?

b) State the energy change in the cooker element.

**8** Name three electrical appliances that change electrical energy into heat.

**9** State two practical uses of a variable resistor.

**10** Draw the circuit symbol for (a) a switch, (b) a lamp, (c) an ammeter, (d) a resistor, (e) a voltmeter, (f) a variable resistor.

# Mains electricity

## After studying this section you should be able to...

1 State that household wiring connects appliances in parallel so that they receive the same voltage.
2 State that the declared value for mains voltage is 230 V.
3 Draw and identify the circuit symbol for a fuse.
4 Identify live, neutral and earth wires from the colour of their insulation.
5 Identify the live, neutral and earth terminals in a plug.
6 Explain how a fuse acts as a safety device.
7 State that electrical energy costs for the home increase as:
a) the power of the appliance increases,
b) the time of use of the appliance increases.
8 Calculate current using: current = $\frac{\text{power}}{\text{voltage}}$.
9 Use the above relationship to determine fuse values.
10 State that a circuit breaker is an automatic switch which can be used instead of a fuse.
11 State that the human body is a conductor of electricity and that moisture increases its ability to conduct.
12 State that the earth wire is a safety device.
13 State that electrical appliances, which have the double insulation symbol, do not require an earth wire.
14 Explain why connecting too many appliances to one socket is dangerous.
15 Explain why situations involving electricity could result in accidents.
16 Describe how to make a simple continuity tester.
17 Describe how a continuity tester may be used to identify an open circuit.

### Power

When an electric current passes through a wire some of the electrical energy is changed into heat in the wire. The amount of heat produced depends on the current in the wire and the resistance of the wire. The wire in an electric fire or kettle where the heat is produced is called the element. Heating elements for electric fires and kettles change electrical energy into heat energy in the resistance wire inside the element. A lamp transfers electrical energy into heat and light in a resistance wire called the filament. How quickly it does this is known as the power rating or power of the lamp. Power is measured in units called watts (W).

## Current, power and voltage

Four different lamps of known power ratings were connected to an electrical supply; the readings obtained for the voltage across and the current through the lamps are shown in table 2.2.

| Current in amperes | Voltage in volts | Power in watts | Power/voltage |
|---|---|---|---|
| 0.5 | 12 | 6 | 0.5 |
| 2.0 | 12 | 24 | 2.0 |
| 3.0 | 12 | 36 | 3.0 |
| 4.0 | 12 | 48 | 4.0 |

**Table 2.2**

Notice that the value of current is the same as the power divided by the voltage for each lamp.

From this we have:

$$\text{current} = \frac{\text{power}}{\text{voltage}}$$

$$I = \frac{P}{V}.$$

### Examples

The interior light of a car is operated from a 12 volt car battery. The power rating of the lamp is 3 watts when it is switched on. Calculate the current in the lamp.

Solution

$$I = \frac{P}{V} = \frac{3}{12} = 0.25 \text{ amperes.}$$

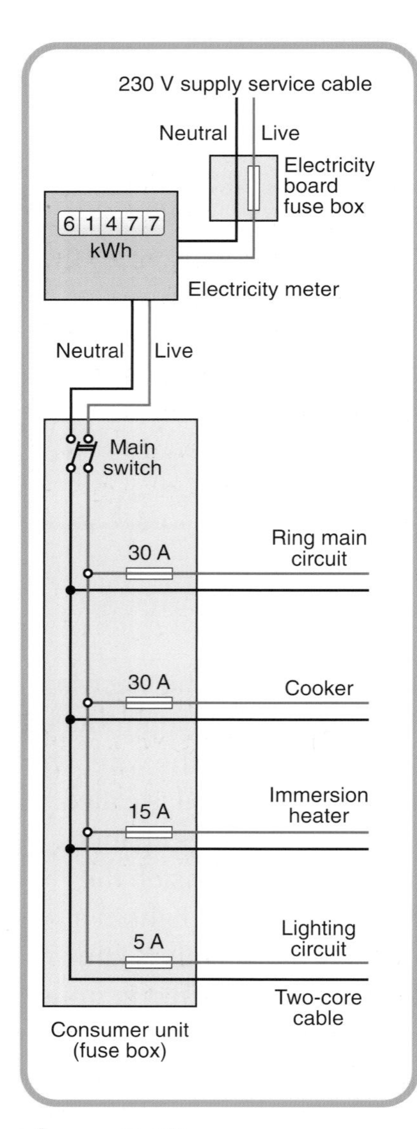


**Figure 2.33**

## Electricity and house wiring

Electricity arrives in our homes by a cable called the service cable. The electricity board fuse box, electricity meter and the consumer unit (sometimes called the fuse box) are connected to it as shown in figure 2.33. The domestic circuits — the lighting and ring main circuits (the sockets in a room) — are connected to the consumer unit. The lighting and ring main circuits are connected in parallel to the consumer unit. This means that the lights and appliances connected to these circuits have the same voltage across them. It also means that when one circuit is switched off the other circuits will still be on. The voltage required to operate household mains appliances is 230 volts (230 V) — this is also known as the declared value for mains voltage.

## Mains fuses and circuit breakers

Domestic wiring circuits are protected by fuses in the consumer unit. They are used to protect the hidden cables (behind the walls) from overheating. The fuse — a thin piece of wire — heats up when an electric current passes through it. When the current is too high the fuse will melt and break the electrical circuit. The circuit symbol for a fuse is shown in figure 2.34.

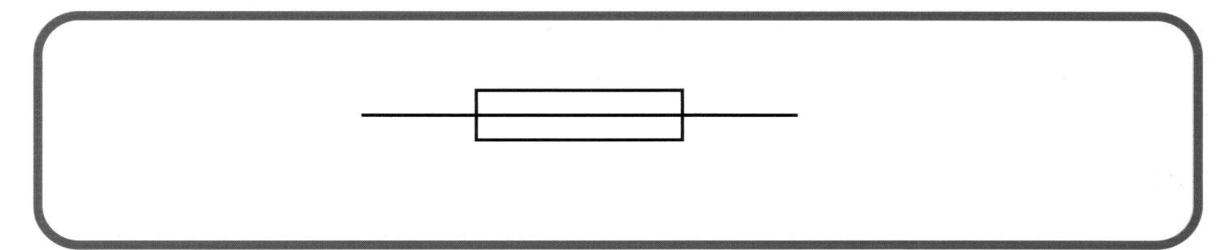

**Figure 2.34**

## Circuit breakers

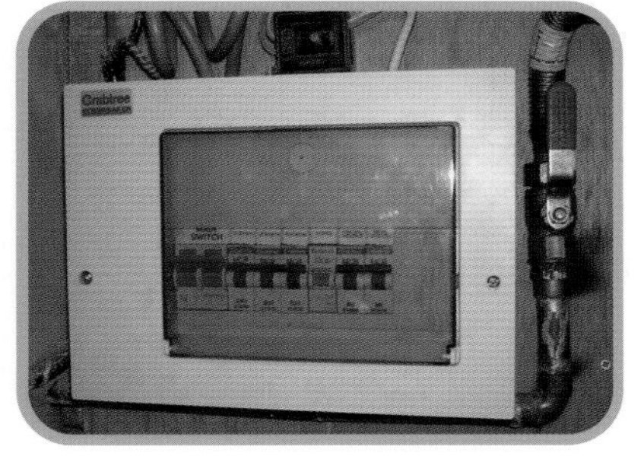

**Figure 2.35**

Figure 2.35 shows a consumer unit fitted with circuit breakers. Circuit breakers are used in some consumer units instead of fuses. These automatically switch themselves off (trip) if there is too much current in the circuit. When the fault has been corrected, the circuit can be reconnected simply by resetting the circuit breaker. A fuse, however, could be replaced with a higher value of fuse wire, allowing a faulty circuit to work. This situation could cause the cables to overheat and catch fire. This cannot happen if a circuit breaker is used as it will continue to trip until the fault has been corrected. Before replacing a fuse or resetting a circuit breaker, the fault should be found and corrected.

## Flexes, plugs and fuses

**Figure 2.36**

Household appliances are connected to three-pin sockets using a flexible cord (or flex for short). A flex consists of two or three cores of thin, copper wire covered with insulation as shown in figure 2.36. The brown covered core is the live wire, the blue covered core is the neutral wire and, in three-core flexes, the green and yellow striped covered core is the earth wire.

The cores in a flex during normal use heat up when a current passes through them. However, if they carry too high a current they will become too hot and overheat, and this could cause a fire. To protect the flex from too high a current, the three-pin plug connected to the flex is fitted with a fuse. A fuse is, simply, a thin piece of wire. When an electric current passes through it, it heats up. If the current gets too high, then the fuse wire will become so hot that it melts or 'blows', breaking the electrical circuit and so protecting the flex (and the appliance).

Fuses for three-pin plugs, known as cartridge fuses, are available in a number of sizes — the most common being 3 ampere (3 A) and 13 ampere (13 A).

## Choosing a fuse

To select the correct size of fuse the power rating or wattage (W) of the appliance, marked on the rating plate, must be known. The rating plate of a vacuum cleaner is shown in figure 2.37.

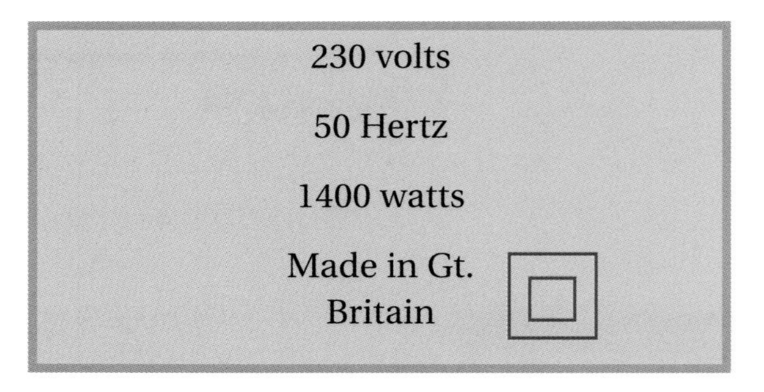


**Figure 2.37** *Rating plate of a vacuum cleaner.*

The current passing through the appliance can then be calculated using the equation $I = \frac{P}{V}$.

The fuse value is required to be larger than the calculated current but closest to the commonly used cartridge fuse values of 3 A and 13 A.

Table 2.3 shows the power ratings and cartridge fuse values for four household (230 volt) appliances.

| Appliance | Power rating in watts | $I = \frac{P}{V}$ | Fuse fitted in plug |
|---|---|---|---|
| Food mixer | 350 | 1.5 amperes | 3 A |
| Kettle | 2200 | 9.6 amperes | 13 A |
| Vacuum cleaner | 1200 | 5.2 amperes | 13 A |
| Toaster | 950 | 4.1 amperes | 13 A |

**Table 2.3**

In general, if the power rating is 700 watts or less then a 3 A fuse should be fitted; if the power rating is greater than 700 watts fit a 13 A fuse.

## Wiring a three-pin plug

For safety, it is very important that a three-pin plug is correctly wired (figure 2.38). Provided the appliance is working properly, then electrical current only passes through the live and neutral cores of the flex — no current passes through the earth wire. The earth wire is connected to the outer metal casing of the appliance and to the earth pin in the three-pin plug. Its purpose is to provide a very easy path for electrical current to pass to earth should a fault occur and it may be considered as a 'safety' wire.

However, if the appliance develops an electrical fault, such as the one shown in figures 2.39 and 2.40 then the earth wire is very important. In figure 2.39 a large current will pass through the live and earth wires. This current will be much larger than the fuse value and so the fuse will 'blow', breaking the electrical circuit. The toaster is now safe for anyone touching it as the broken fuse has disconnected it from the live wire.

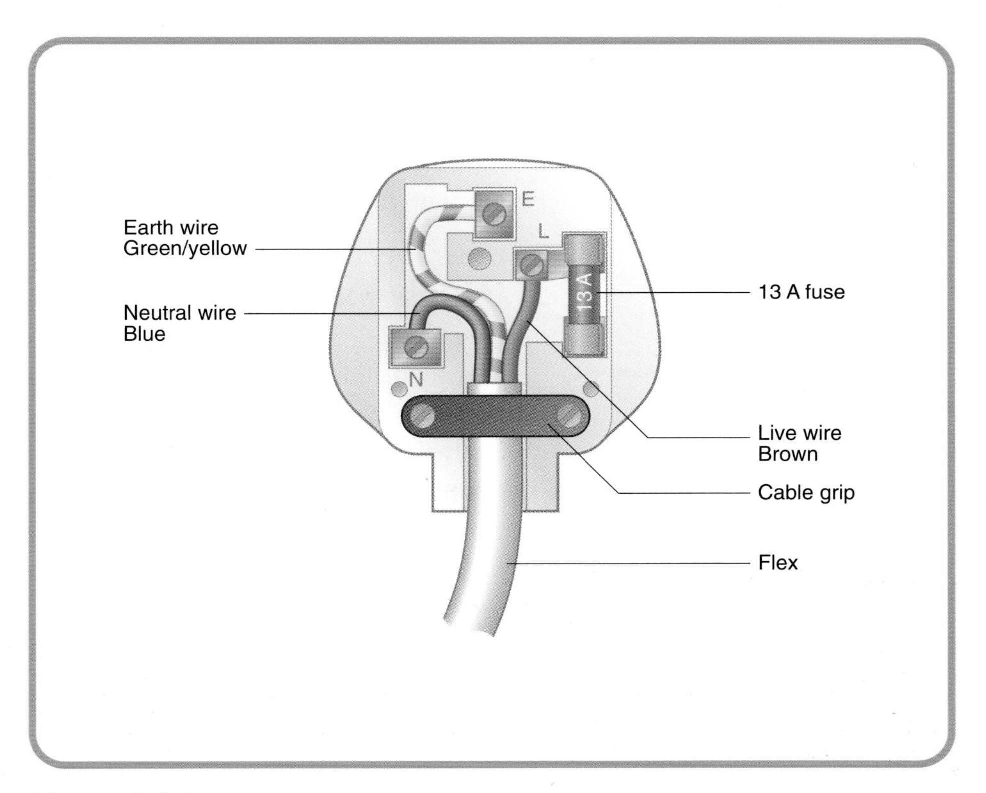


**Figure 2.38**

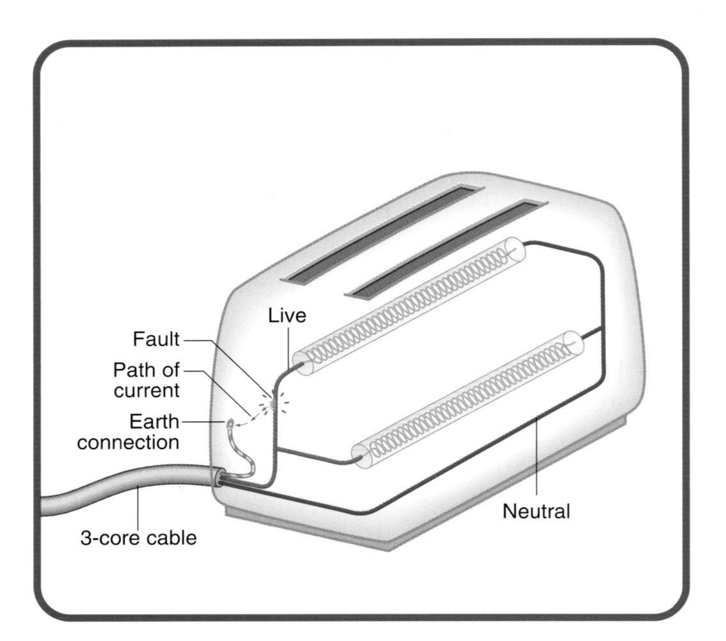


**Figure 2.39**

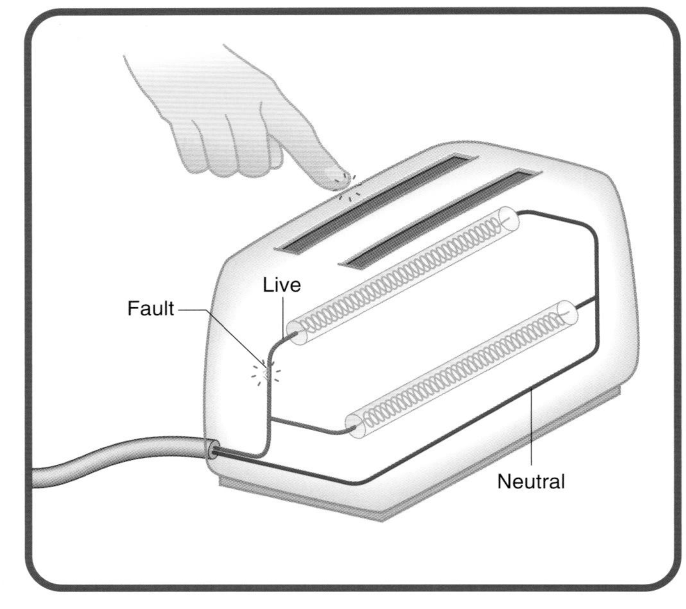


**Figure 2.40**

In figure 2.40 the Earth wire is not connected and so the metal parts of the toaster are 'live'. Anyone touching the toaster will get a shock or possibly be killed — current will pass from the live wire through the person to Earth. The danger is greatly increased if the person's fingers are wet, as water helps to conduct the electricity through your body.

## Double insulation

Some appliances, such as hairdryers and electric drills, do not require an Earth wire since there are two layers of insulation around their electrical parts. This makes the Earth wire unnecessary and so a two-core flex is fitted. The double insulation symbol is shown in figure 2.41. Appliances which are double insulated have this symbol shown on their rating plate. The rating plate of the vacuum cleaner shown in figure 2.37 has the double insulated symbol and this indicates that the vacuum cleaner is double insulated.

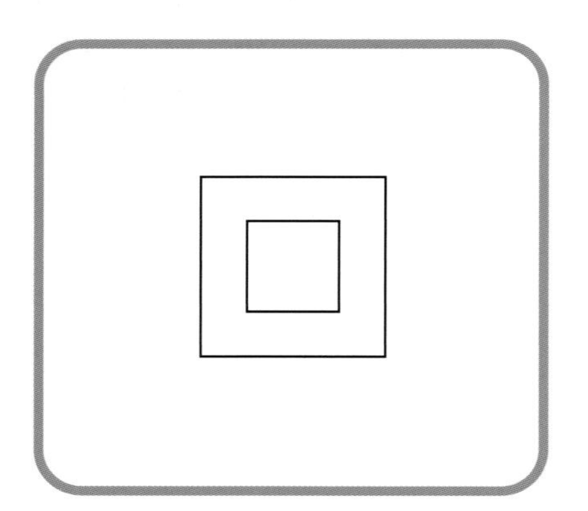

**Figure 2.41** *The double insulation symbol*

## Dangerous situations

The wall socket shown in figure 2.42 has too many appliances plugged into it. This could result in too much current being drawn from the wall socket, which could lead to the socket overheating and causing a fire.

**Figure 2.42**

When your body is damp or wet, you become a better conductor of electricity. Touching wall sockets or light switches with wet or damp hands can be dangerous as the water makes you a better conductor. If water entered the socket or light switch this could provide a path for the current to pass through your body and you could be electrocuted. It is for this reason that bathrooms do not have wall sockets and the light switch is either fitted with a cord inside the bathroom or placed on a wall outside the bathroom.

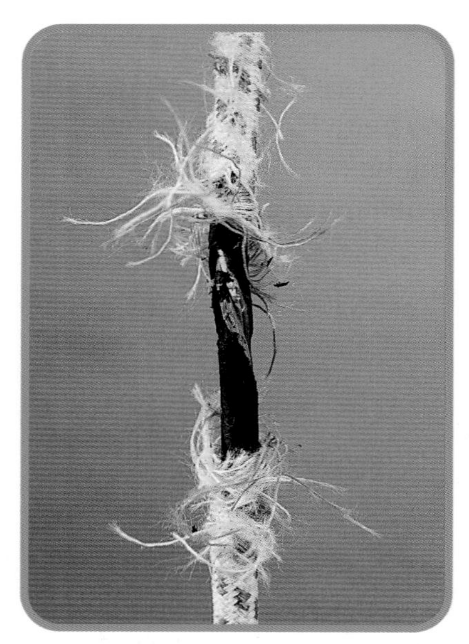

Figure 2.43

Frayed, worn or joined flexes are all dangerous (figure 2.43) since the live core may become bare and someone may touch it – resulting in a 'shock' or electrocution. In these cases a replacement flex of the correct rating must be fitted.

## Calculating your electricity bill

The electricity meter measures the electrical energy used by the appliances in your house.

The more units of electricity that are used in your house, the higher the cost of your electricity bill. The cost of using an electrical appliance depends on:

- the power rating of the appliance — the higher the power rating, the greater the cost,
- the time the appliance is used for — the longer the appliance is on, the greater the cost.

If all the appliances in table 2.3 were switched on for the same length of time, then the kettle would cost the most (it has the highest power rating) and the food mixer would cost the least (the smallest power rating).

### Examples

The element of a cooker has a power rating of 2000 watts. It costs 12 p to use the element when switched on for 1 hour. How much will it cost to use the cooker element for (a) 2 hours, (b) $\frac{1}{2}$ hour?

Solution

a) The cooker element costs 12 p for 1 hour. As the element is switched on for 2 hours, twice the time of 1 hour, it will cost twice as much to use:
Cost = 2 x 12 p = 24 p

b) As the element is switched on for $\frac{1}{2}$ hour, half the time of 1 hour, it will cost half as much to use:
Cost = 0.5 x 12 p = 6 p

## Examples

It costs 6 p to have an electric fire on for 1 hour. How many hours could this fire be used for if the cost is (a) 12 p, (b) 36 p?

Solution

a) The fire costs 6 p for 1 hour; 12 p is twice the cost. The fire is on for twice as long, so the fire could be used for 2 hours.

b) Fire costs 6 p for 1 hour; 36 p is six times the cost. The fire is on for six times as long, so the fire could be used for 6 hours.

## A simple circuit tester

A circuit tester is used to test if a wire or component is broken. A cheap and simple circuit tester can be made from a battery, lamp, resistor and some wires as shown in figure 2.44. The resistor R is present so that when wires X and Y are joined, the lamp is as bright as it safely can be. R is called a protective resistance and prevents too much current passing through the lamp and breaking it. Any other piece of electrical equipment, for example a resistor or a lamp joined between X and Y, will increase the resistance of the circuit. There will be less current in the circuit and the lamp will be less bright.

Before using the tester to detect for a fault, X and Y are joined together to check that the lamp lights. If it does not light, the lamp or battery should be replaced — after having checked for a loose connection.

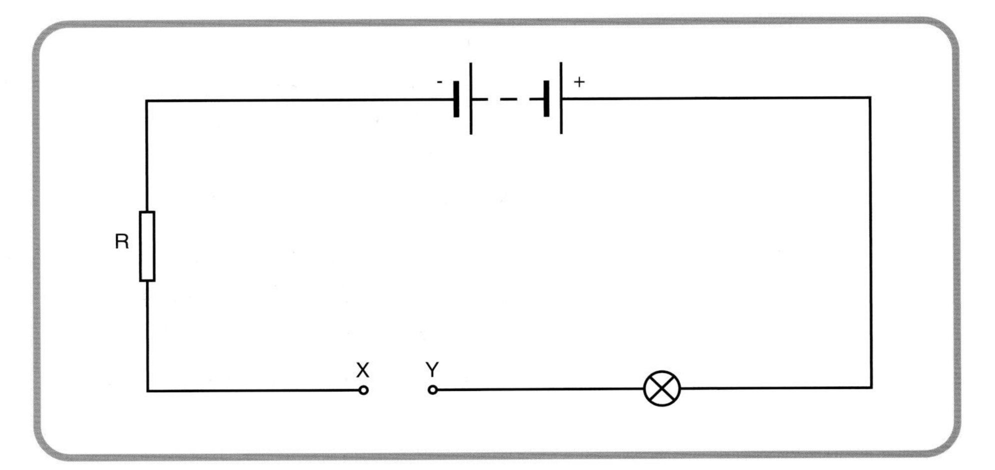


**Figure 2.44**

## Fault finding

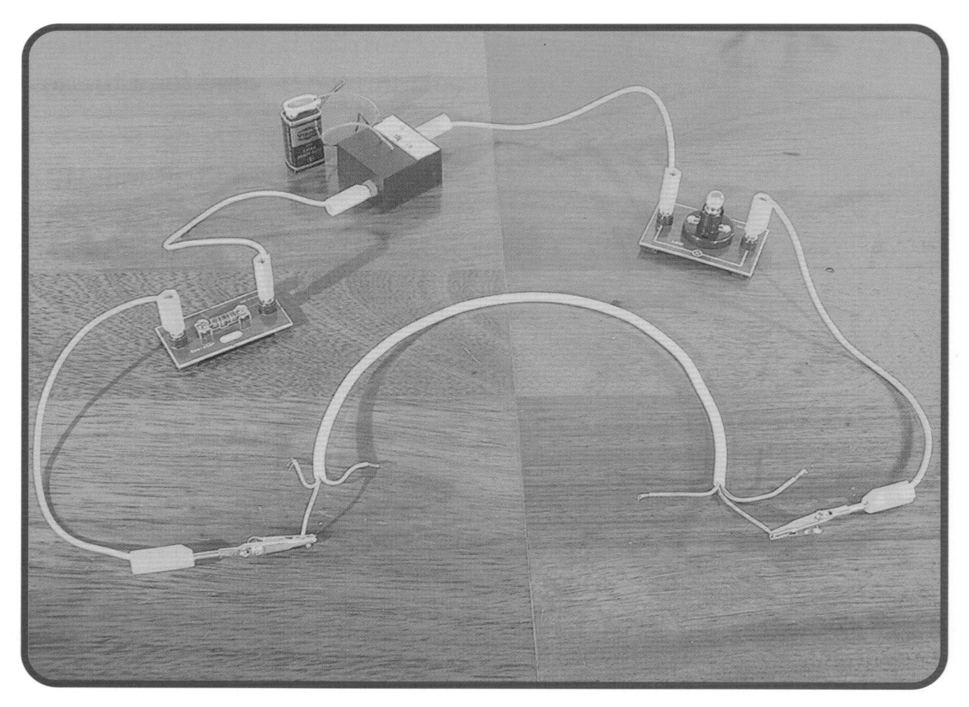

**Figure 2.45**

The component or components must be disconnected from the power supply before using the continuity tester to test for a fault. Figure 2.45 shows a wire with an electrical break connected to the circuit tester. The lamp does not light. This is known as an open circuit. Open circuits can occur with both electrical components and wires.

Consider a circuit in which a lamp does not light. This could be due to:

- the filament of the lamp being broken,
- a broken wire between the battery and the lamp,
- the battery being 'flat'.

Using the simple circuit tester, the fault in this circuit may be found by:

a) Testing the lamp:
   - Connect the lamp to the continuity tester – if the lamp lights, the lamp filament is not broken.
   - If the lamp does not light – the lamp filament is broken (fault found).

b) Testing the wires in the circuit. Connect each wire to the continuity tester as shown in figure 2.45:
   - If the lamp does light –the wire is not broken.
   - If the lamp does not light –the wire is broken (fault found).

If none of the above has detected a fault then the battery should be replaced.

## Key Points

- Voltage is measured in volts (V), current in amperes (A), resistance in ohms (Ω) and power in watts (W).
- Current = $\frac{\text{power}}{\text{voltage}}$.
- In a lamp, electrical energy is changed into heat and light in the resistance wire (filament).
- Appliances are connected in parallel to the household wiring.
- Mains fuses protect the mains wiring from overheating.
- Circuit breakers, a type of automatic switch, can be used instead of fuses.
- Circuit breakers are safer than fuses since fuses can be replaced with a wrong (higher) value of fuse.
- Household electricity meters measure the electrical energy used in a house.
- The cost of using an electrical appliance depends on the power rating of the appliance and the length of time the appliance is used: (a) the cost increases as the power rating increases; (b) the cost increases as the time the appliance is used increases.
- Appliances with a power rating up to 700 watts are normally fitted with 3 ampere (3 A) fuses.
- Appliances with a power rating greater than 700 watts are normally fitted with 13 ampere (13 A) fuses.
- Fuses are intended to protect the flex of an appliance from overheating.
- Three-core flexes consist of live (brown), neutral (blue) and Earth (green and yellow) wires.
- Two-core flexes, used with double insulated appliances, consist of live (brown) and neutral (blue) wires.
- The Earth wire is a safety device.
- A simple continuity tester can be made from a battery, lamp, resistor and wires.
- A simple continuity tester can be used to find an open circuit — a broken wire or component (lamp will not light).

## Section Questions

**1** Name the components shown in figure 2.46.

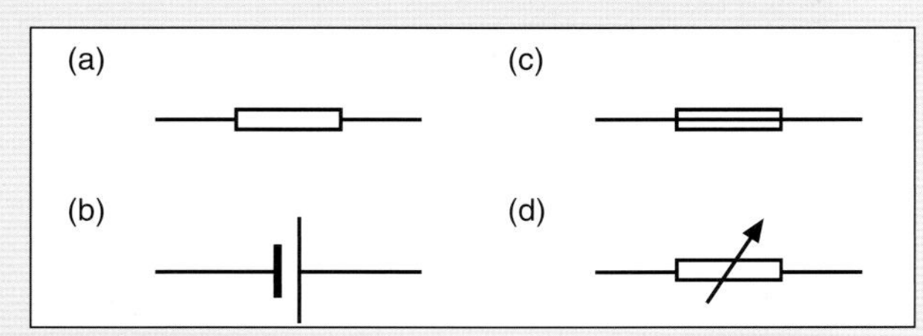


**Figure 2.46**

**2** Copy and complete the following sentences using the appropriate words from the list: declared, parallel, voltage.

Household wiring connects appliances in ______. This means each appliance receives the same ______. The ______ value for mains voltage is 230 volts.

**3** Copy and complete the following sentences using the appropriate words from the list: conduct, high, long, low, poor, short.

The human body is normally a ______ conductor of electricity. However, its ability to ______ electricity is increased when it is damp. It costs more to use an appliance with a ______ power rating than one with a ______ power rating for the same time. It costs more to use an appliance for a ______ time than a ______ time.

**4** Figure 2.47 shows a flex connected to a three-pin plug.

a) Name pins A, B and C.

b) State the colours of the insulation on the wires P, Q and R.

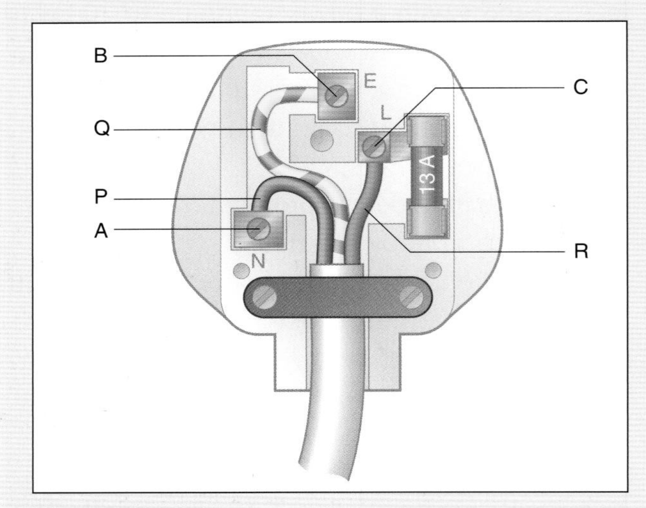


**Figure 2.47**

**5** Copy and complete table 2.4.

| | Current in amperes | Power in watts | Voltage in volts |
|---|---|---|---|
| (a) | | 24 | 12 |
| (b) | | 690 | 230 |
| (c) | | 2300 | 230 |

**Table 2.4**

**6** The power rating of a lamp is 48 watts. The lamp is connected to a 12 volt supply. Calculate the current in the lamp when it is working at its correct power rating.

**7** Some information from the rating plates of a games console and a hairdryer is given below.

| 230 volts, 46 watts<br>Games console | 230 volts, 1150 watts<br>Hairdryer |
|---|---|

a) The appliances are working at their stated values. Calculate the current in (i) the games console, (ii) the hairdryer.

b) The plug for each appliance should be fitted with either a 3 ampere or 13 ampere cartridge fuse. Which size of cartridge fuse should be fitted in the plug connected to (i) the games console, (ii) the hairdryer?

**8** The consumer unit in many houses is fitted with automatic switches which take the place of fuses. What is the name of these automatic switches?

**9** Explain why it could be dangerous to connect four appliances into the same wall socket using adapters.

**10** Explain the danger involving the use of an electrical appliance:

a) with a frayed flex,

b) near a sink of water (see figure 2.48).

a)

b)

**Figure 2.48**

**11** Draw the circuit symbol for (a) a variable resistor, (b) a fuse, (c) an ammeter, (d) a resistor, (e) a voltmeter.

## Examination Style Questions

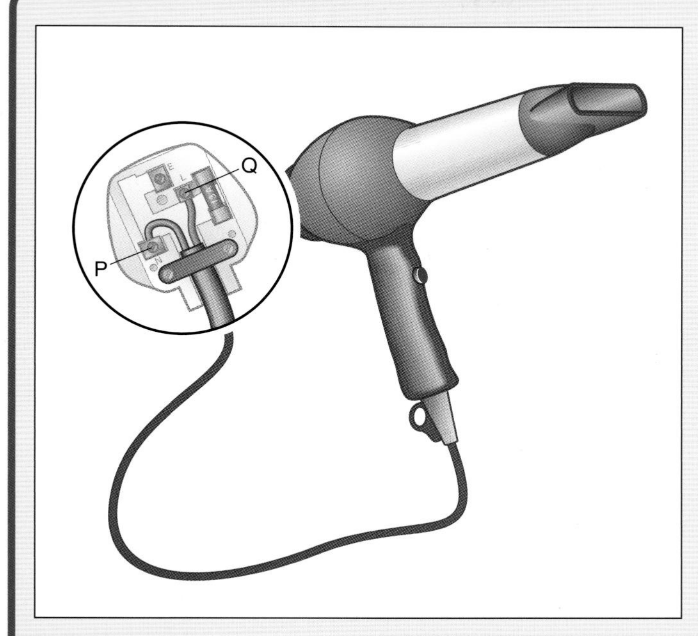


**Figure 2.49**

**1** Figure 2.49 shows a hairdryer correctly connected to a three pin plug. The hairdryer is double insulated.

a) (i) Draw the double insulation symbol.
   (ii) Name the pin in the plug which has not been used.

b) Name the colour of the insulation on the wire connected to:
   (i) pin P,
   (ii) pin Q.

c) The hairdryer has a power rating of 1035 watts.
   (i) Calculate the current in the hairdryer when it is connected to the 230 volt mains supply.
   (ii) The following fuses are available:
   3 ampere  5 ampere  13 ampere

   Which fuse should be fitted in the three pin plug connected to the hairdryer?

**2** A lamp is connected in series with a variable resistor to a 9 volt supply as shown in figure 2.50. A voltmeter is added to measure the voltage across the lamp as shown.

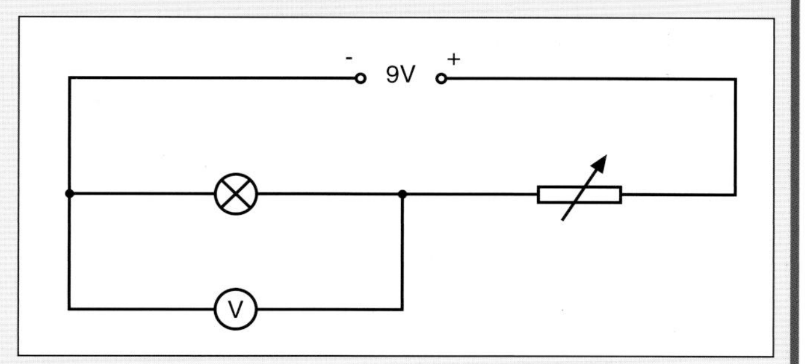


**Figure 2.50**

a) Copy figure 2.50. Complete your diagram to show how an ammeter is connected to measure the current in the lamp.

b) The reading on the voltmeter is 3.5 volts. The current in the lamp is 0.2 amperes. Calculate the resistance of the lamp.

c) The resistance of the variable resistor is decreased.
   (i) What happens to the brightness of the lamp?
   (ii) Explain your answer.

**3** A circuit is set up as shown in figure 2.51.

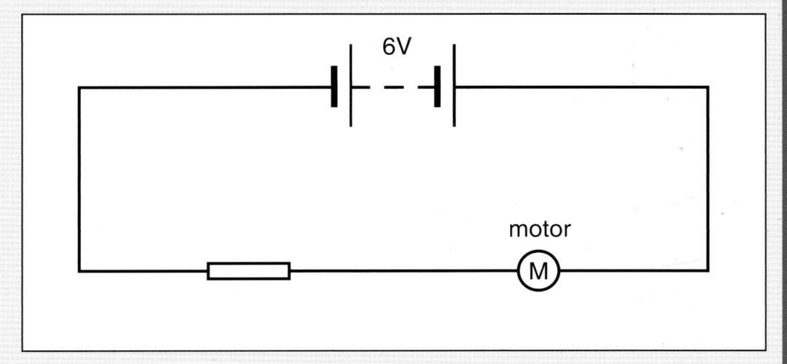


**Figure 2.51**

a) Copy figure 2.51. Complete your diagram to show how a voltmeter is connected to measure the voltage across the motor.

b) The voltage across the motor is 4 volts. What is the voltage across the resistor?

c) The current in the motor is 0.06 amperes:
   (i) What is the current in the resistor?
   (ii) Calculate the resistance of the resistor.

**4** Figure 2.52 shows the connections between a 12 volt car battery, a rear fog light and a rear window heater.

a) Which switch or switches must be closed to allow only the rear fog light to come on?

b) When switched on, the current in the rear window heater is 10 amperes. When switched on, the current in the rear fog lamp is 5 amperes. When the rear window heater and the rear fog lamp are both switched on, what current is supplied by the battery?

c) Fuses for the circuit are available as shown below.

| | | |
|---|---|---|
| 3 amperes | 5 amperes | 10 amperes |
| 13 amperes | 15 amperes | 30 amperes |

Select the most appropriate fuse value for the circuit.

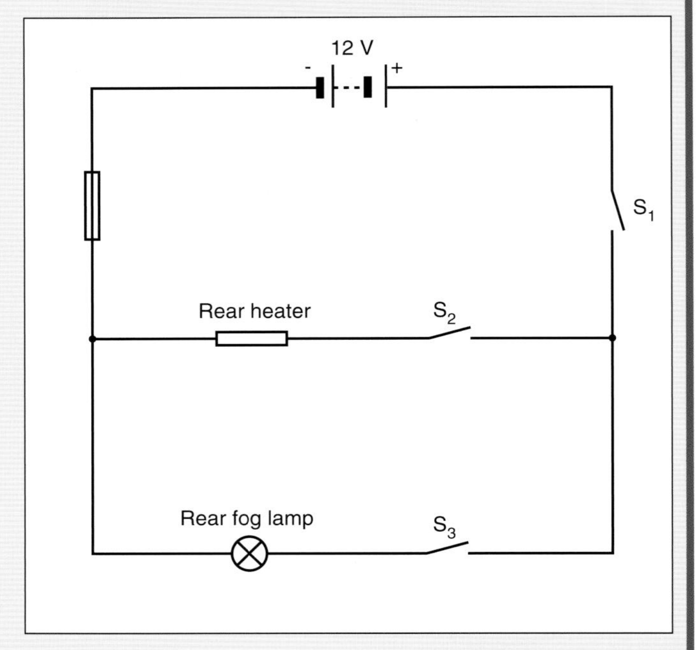


**Figure 2.52**

Unit 3

# Radiations

Different radiations have different uses. Lasers are used in CD and DVD players and medicine. Light lets us see what is happening and infrared is heat radiation; ultraviolet is needed for a suntan but only a small amount. Finally X-ray and radioactive sources can be used to detect and treat a range of diseases.

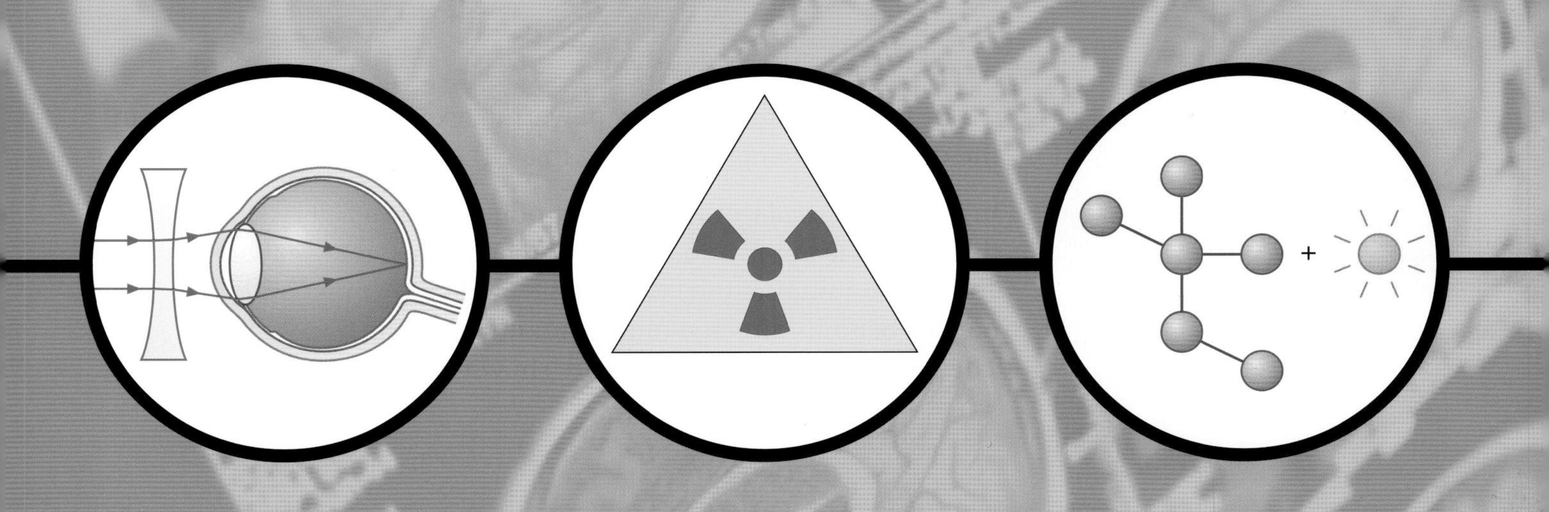

# Lasers

## After studying this section you should be able to...

1. State that a laser is a concentrated source of light of only one colour.
2. Describe how a laser is used in one practical application.
3. State that light can be reflected.
4. State that all visible objects give out, or reflect, light to the eye.
5. Describe the direction of reflected light from a plane mirror.
6. Describe one use of optical fibres in medicine.
7. Describe the shapes of converging and diverging lenses.
8. Describe the effect of converging and diverging lenses on parallel rays of light.
9. Describe, in words, or using a diagram, the eye defects called long and short sight.
10. State that a converging lens can correct long sight and a diverging lens can correct short sight.

**Figure 3.1**

The word laser is made from the initial letters of the words that make it up: **L**ight **A**mplification by **S**timulated **E**mission of **R**adiation.

If you look at the light coming from a lamp it spreads out in all directions. A laser is a very concentrated form of light (figure 3.1). The light travels in a straight line and spreads out very little. This makes the light very concentrated. If this light reaches the eye it can damage it. The light is also of one particular colour. A helium neon laser gives out red light and an argon laser gives out a blue green light (figure 3.2). Two American scientists made the first laser in 1960. It was described as a 'solution looking for a problem'!

## Uses of lasers

**Figure 3.2**

CD or DVD writers use a very powerful laser to burn into the dye in the disc and a less powerful one in a CD or DVD player (figure 3.3).

In medicine, the laser is used to produce heat in a very small piece of tissue. In one application, the laser beam is used to seal blood vessels. To remove tumours the laser can be made into a narrow beam. This narrow beam is directed onto a tumour, causing it to be burnt away. The colours of different lasers are shown in table 3.1.

**Figure 3.3** *Inside a CD player.*

| Laser | Colour |
| --- | --- |
| Carbon dioxide | Infrared, that is heat |
| Helium neon | Red |
| Argon | Green to blue |
| Excimer | Beyond the blue (ultraviolet) |

**Table 3.1**

## Laser as a scalpel

The energy in the carbon dioxide laser beam can be taken in by a small amount of tissue. This idea can be used as a 'laser scalpel'. This has the advantage of there being no blood when a cut is made in the skin in certain operations. Any blood vessels are sealed by the heat produced in the laser when it reaches the tissue. Certain cancerous tumours can be changed into a gas using a carbon dioxide laser.

## Eye problems

Eye surgery is the best known application of argon lasers. The retina of the eye, which is the back part of the eye, sometimes does not get enough oxygen from blood vessels in a diabetic person. To try and overcome this lack of oxygen, there is bleeding into the eye. If nothing is done then the sight at the edge of the eye is reduced and the patient can eventually go blind (figure 3. 4).

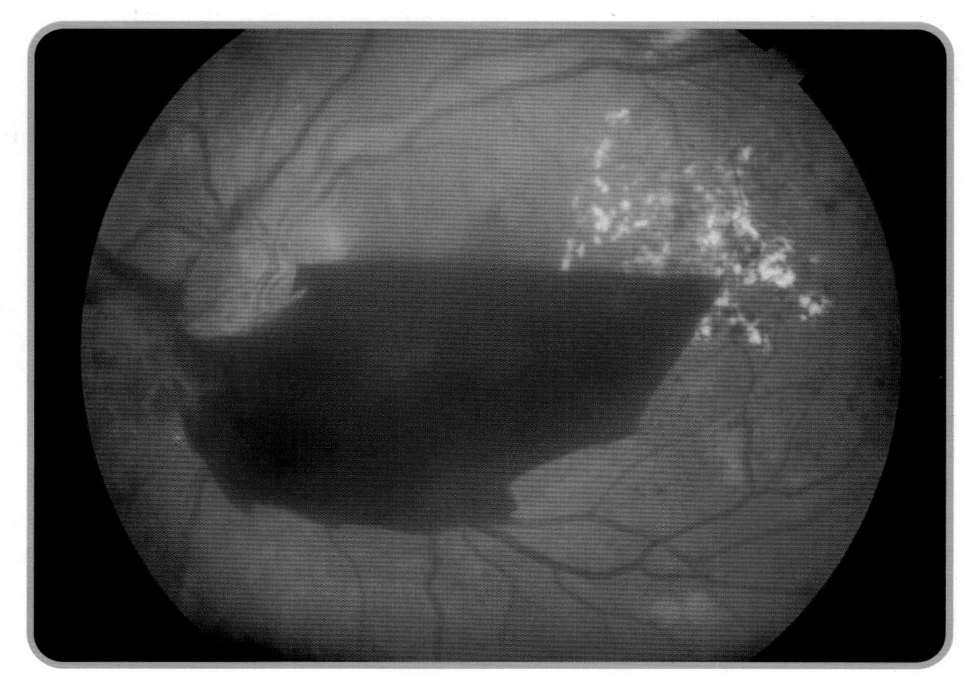

**Figure 3.4** *Damaged eye of a diabetic person.*

The eye surgeon uses an argon laser to seal the less important areas of the retina. Although the patient does not have as wide a range of vision as before, he/she is much less likely to go blind.

This technique can be used for repairing tears in the retina and holes which develop before the retina comes away from the back of the eye.

## Other uses of lasers

Port wine birth marks are caused by blood vessels which have not sealed properly. The light from the argon laser causes the blood vessels to seal. A similar treatment can be used to remove some tattoos (figure 3.5).

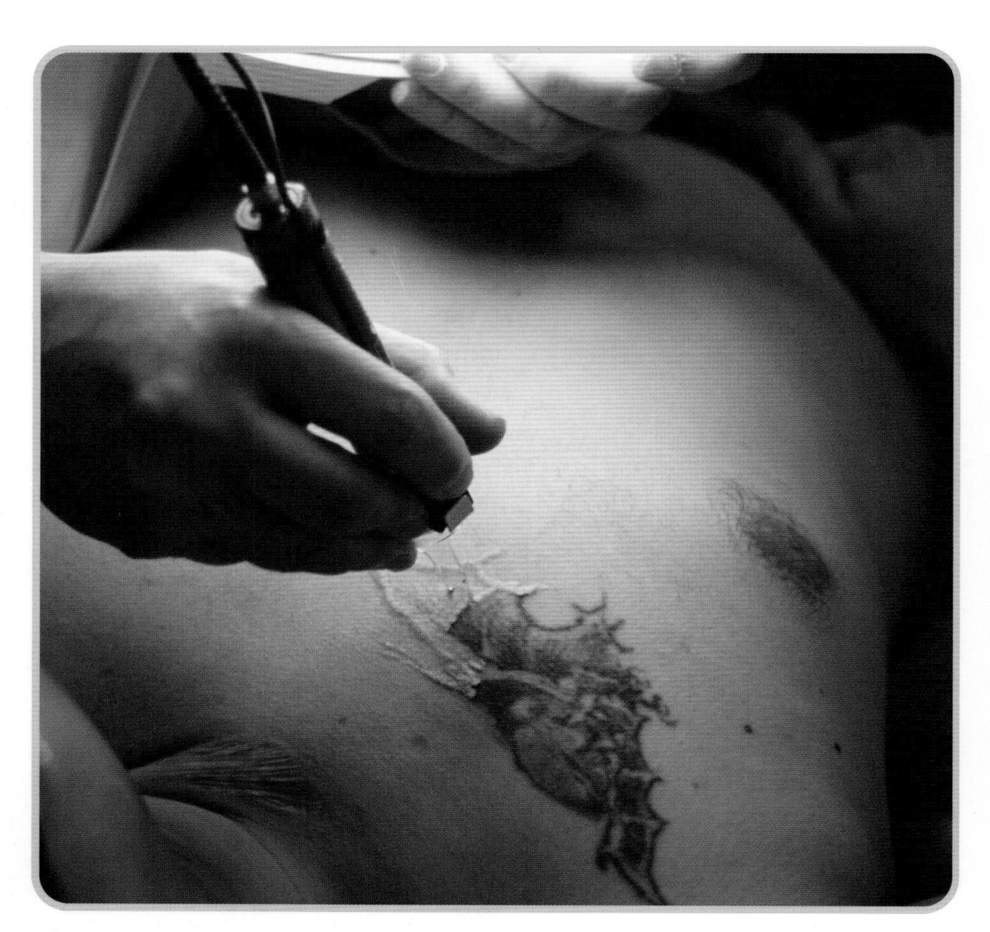

**Figure 3.5**

A further use of lasers is when a patient has had a lens inserted in the eye after a cataract operation. A cataract happens when the eye lens becomes cloudy. The eye lens is removed and replaced by a new plastic lens. Strands of tissue grow behind the eye which do not allow the light to pass through the lens. A few pulses from a laser will split the tissue and restore the patient's sight.

## Physics for you

### Solving eye problems — no glasses needed

Lasers can be used to change the shape of the front of the eye called the cornea. A laser vaporises, that is burns away part of the cornea (figures 3.6 and 3.7). The change is permanent and some patients report problems with their eyes watering or sometimes blurring (figure 3.7).

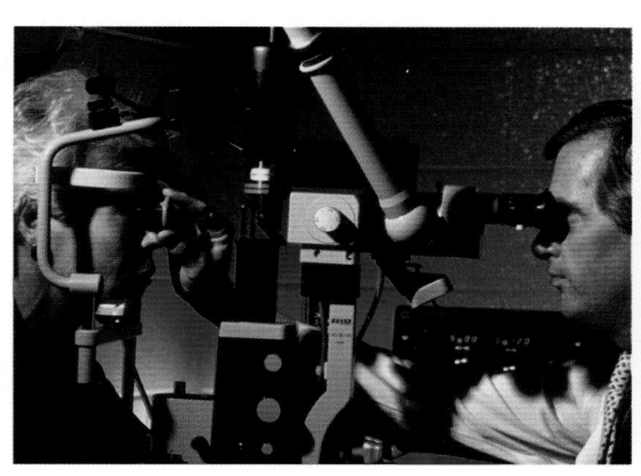

**Figure 3.6** *Laser eye treatment.*

**Figure 3.7**

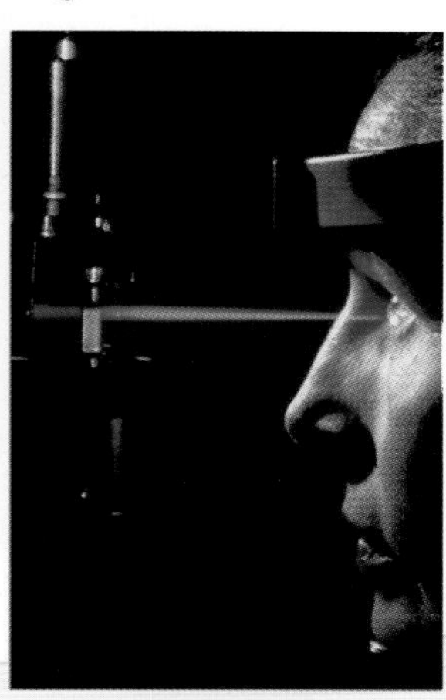

# Goodbye, four eyes

***Civilisation became possible with the invention of spectacles. Now laser surgery is making glasses redundant.***

**The sound of the laser beam incinerating his eyeball did not upset Joe Middleton that much. It was the smell of his cornea being burnt that caused him real distress.**

Given that his laser surgery took only a few minutes, Joe's tribulations were limited. 'Just the same, it was gruelling,' he said.

The experiences of the 37-year-old lawyer are scarcely unique, however, for it is anticipated that 75,000 people will undergo the same unpleasant operation this year, double last year's total.

Despite the squeamish experience of having one's eyes shaped by burning, laser surgery has begun to take off – a fact that was starkly underlined last week when Elton John admitted he, too, would have his eyes scorched and shaved to restore his vision, thus curtailing the need to wear his trademark spectacles.

The experiences of Joe Middleton, a typical recent convert, provide a revealing illustration of the causes of this visual revolution. 'I am short-sighted but have always found both glasses and contact lenses a nuisance,' he said. 'Laser surgery seemed the obvious answer, but I waited for years to see if there were any reports of early patients suffering side-effects before going ahead.'

Yet civilisation probably would never have spread with its spectacular speed had it not been for glasses. The passing of knowledge required the invention of the printing press but it also needed the development of visual aids for reading words that would otherwise have been lost to the blurred, unfocused vision that afflicts our species.

It was not until the late thirteenth century that the first spectacles were developed, though the identity of their inventor has been lost in the myopic mists of time.

Since then, spectacles have been created in an increasingly and bewildering array of types: glass replaced quartz, then plastic replaced glass. We also have bifocals, trifocals, varifocals and of course contact lenses – essentially tiny pairs of spectacles placed directly over the eyes.

**Robin McKie** Sunday November 17, 2002
*The Observer*

## Holograms

Holograms are three-dimensional pictures and were first suggested in 1948. It was not possible to make them until the invention of the laser. A popular use of a special type of hologram called a reflection hologram is used in credit cards for security purposes.

## Reflection of light

The sun or a lamp gives out light which we detect using our eyes. Solid objects reflect light which allows us to see the object. The reflection of light happens because the light is made of several rays, like lines, which reflect off the object (figure 3.8).

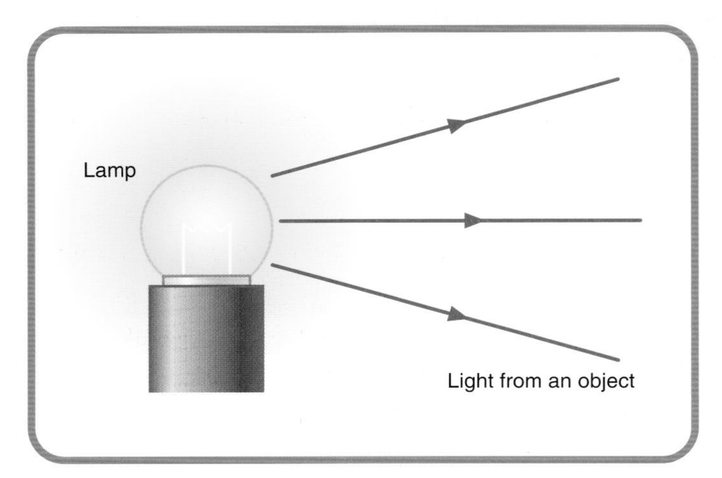


**Figure 3.8**

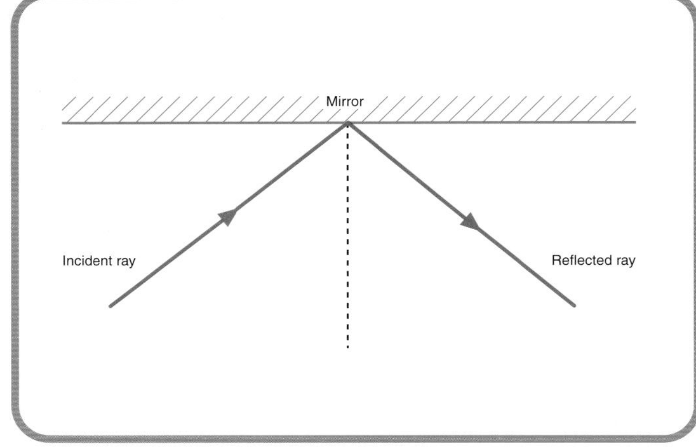


**Figure 3.9**

If a plane mirror is used then the light reflects off so that the angles shown in figure 3.9 are equal.

The ray striking the mirror is called the incident ray and the ray reflected off the mirror is called the reflected ray. The dotted line at right angles is called a normal, and this is the line from which all angles are measured.

If the angle of the incident ray is increased then the angle of the reflected ray increases by the same amount (figure 3.10).

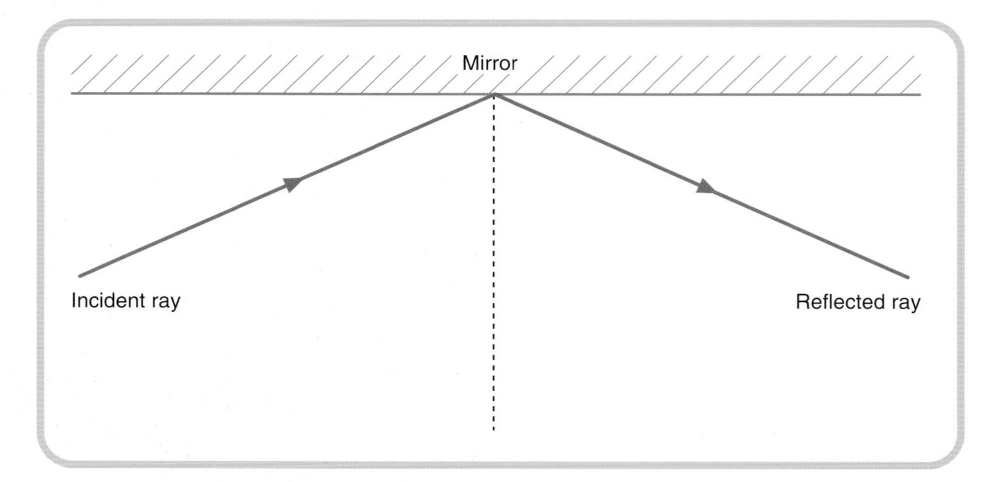


**Figure 3.10**

## Seeing round the bend in medicine

Fibre optics are used in medicine to see inside the body without surgery.

The optical fibres are about the thickness of a human hair. Each fibre consists of a thin piece of glass coated with a thin layer or cladding of another glass. This cladding prevents the light, which enters the end of the fibre, from escaping or passing through the sides to another fibre in the bundle (see unit 1).

Optical fibres in medicine are used in a device called an endoscope or bronchoscope.

The key parts are :

- A bundle of fibres to send light down to the internal organs using total internal reflection.
- Light is then reflected from the organs and an image sent up another bundle of fibres to a camera or the doctor's eye.
- A channel or hollow tube to clean the lens (figure 3.11).

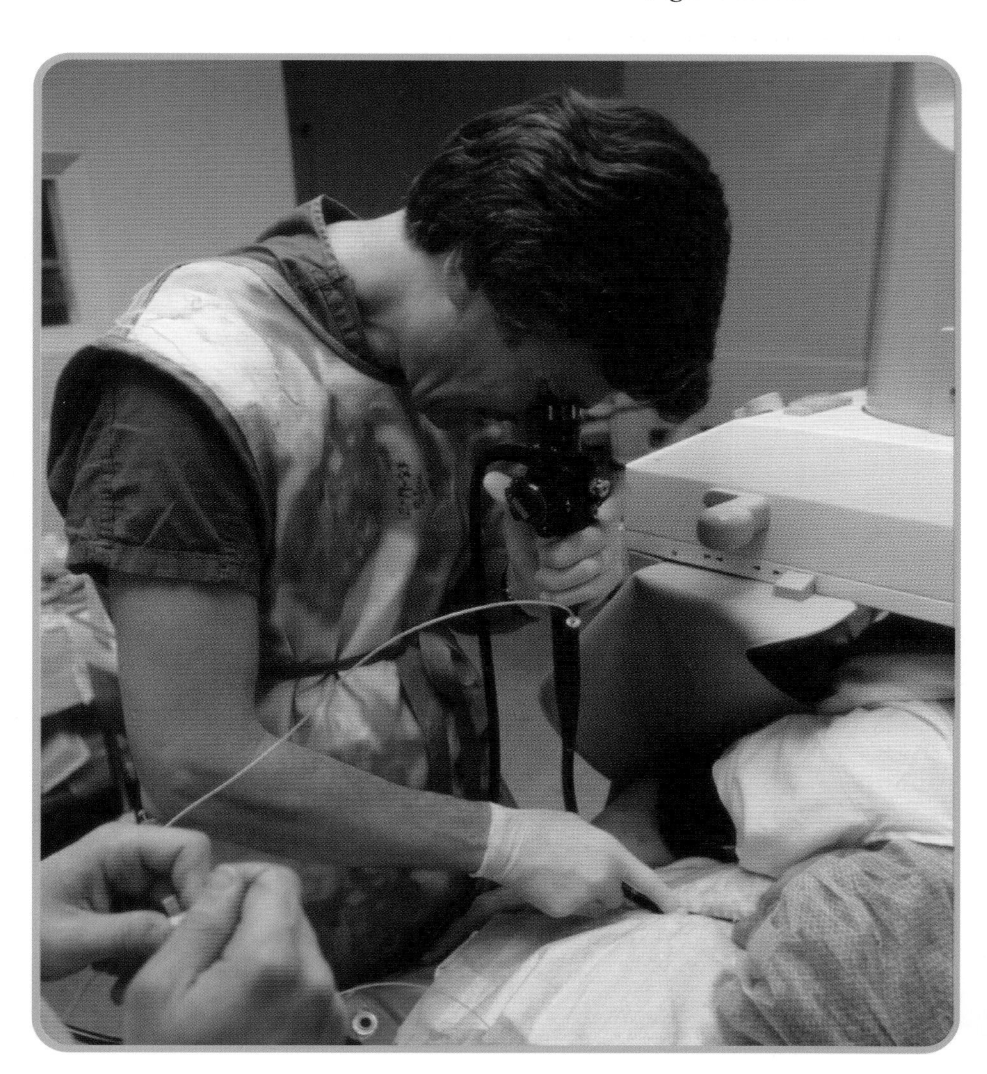

**Figure 3.11** *An endoscope being inserted into a patient.*

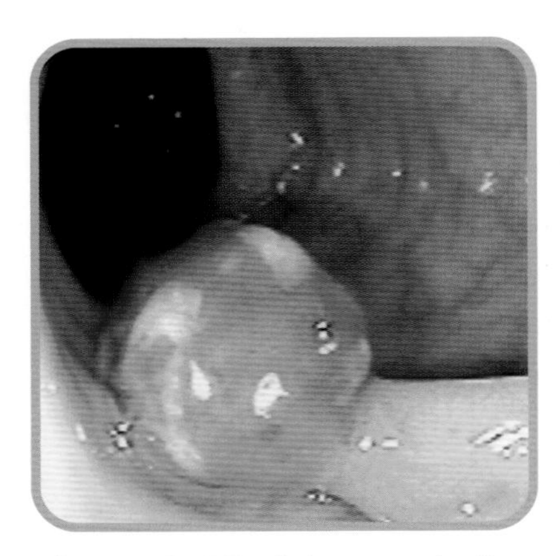

**Figure 3.12** *A tumour in the colon.*

The heat from the lamp does not pass down the fibres. This means that the other end of the guide is cold (called a cold light source). This is one of the advantages of the endoscope.

They have a bending section near the tip so the doctor can direct the instrument during insertion.

In recent operations the endoscope has been used for keyhole surgery. Keyhole surgery means making a small opening in the body to insert the endoscope and small instruments to perform the intricate surgery. The advantage is that the small cut means a faster recovery.

Endoscopes can also be used to view a tumour which might occur in the colon (figure 3.12).

## Lenses

### Physics for you

If you open your eyes underwater, can you see out of the water?

If a diver looks straight out of the water in a lake he will see the sky. If he looks out of the water at a large angle then he will see a reflection of the bottom of the lake (figure 3.13).

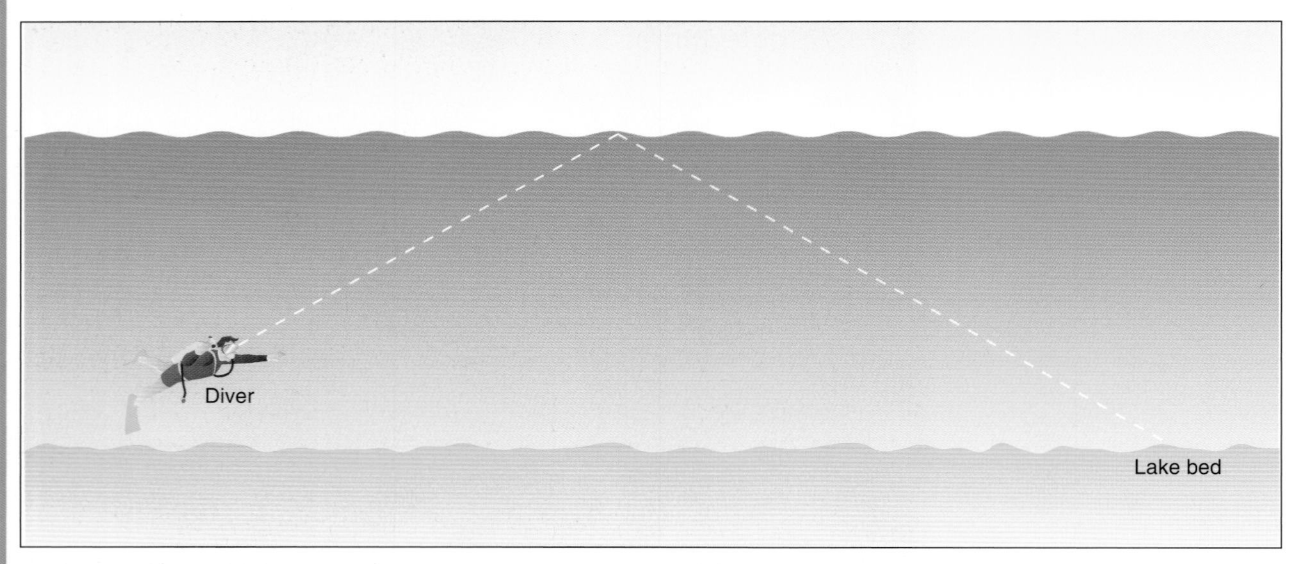


**Figure 3.13**

These are different shapes of glass. There are two types of lenses:

### Convex or converging lens

The middle ray goes straight on and the outer rays bend and meet on the middle line at a point called the focus (figure 3.14). If the lens is thick, the same effect occurs but the focus is nearer the lens. The distance from the lens to the focus is called the focal length.

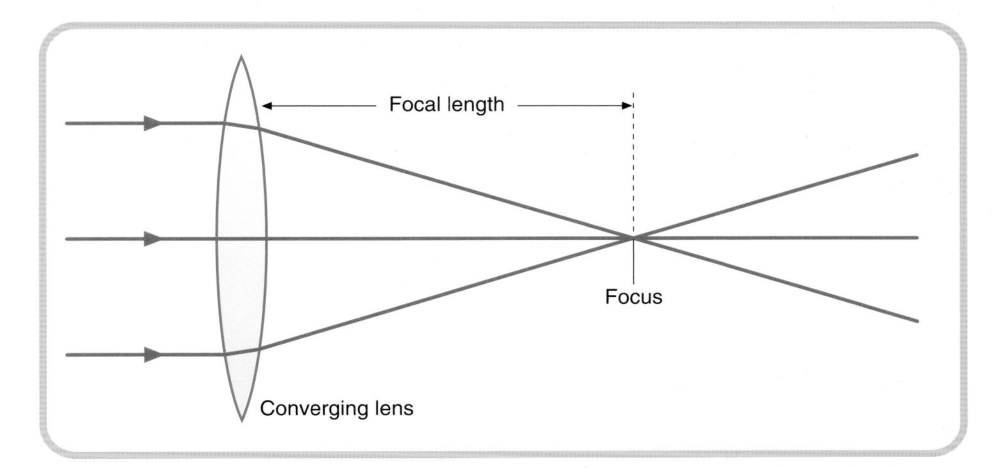


**Figure 3.14**

### Concave or diverging lens

The middle ray goes straight on and the outer rays bend away from the lens. This means that the rays spread out (figure 3.15).

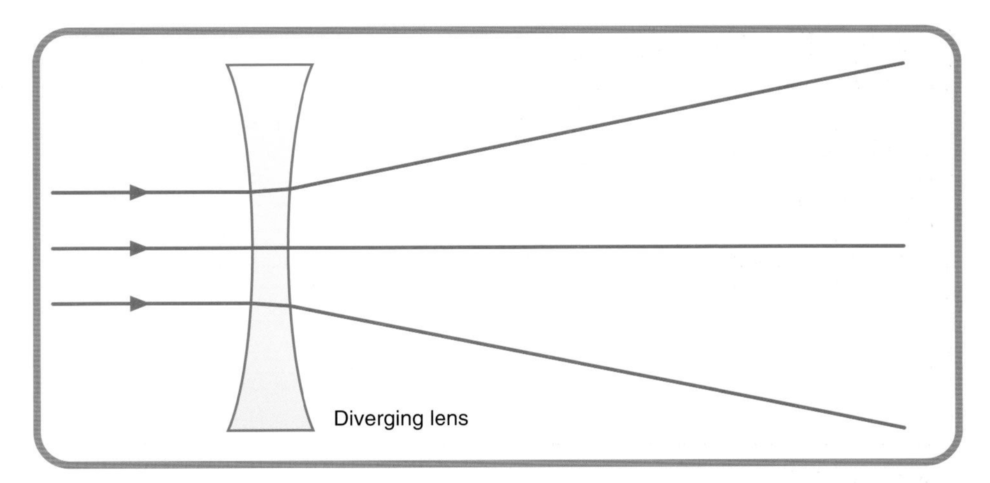


**Figure 3.15**

## The eye

Our eyes tell us what's going on in the 'outside world'. They enable us to grasp, hit or touch objects 'out there' and, of course, to avoid being grasped, hit or touched by threatening objects. Our eyes tell us the shape, size and colour of objects. An outline of the eye is shown in figure 3.16.

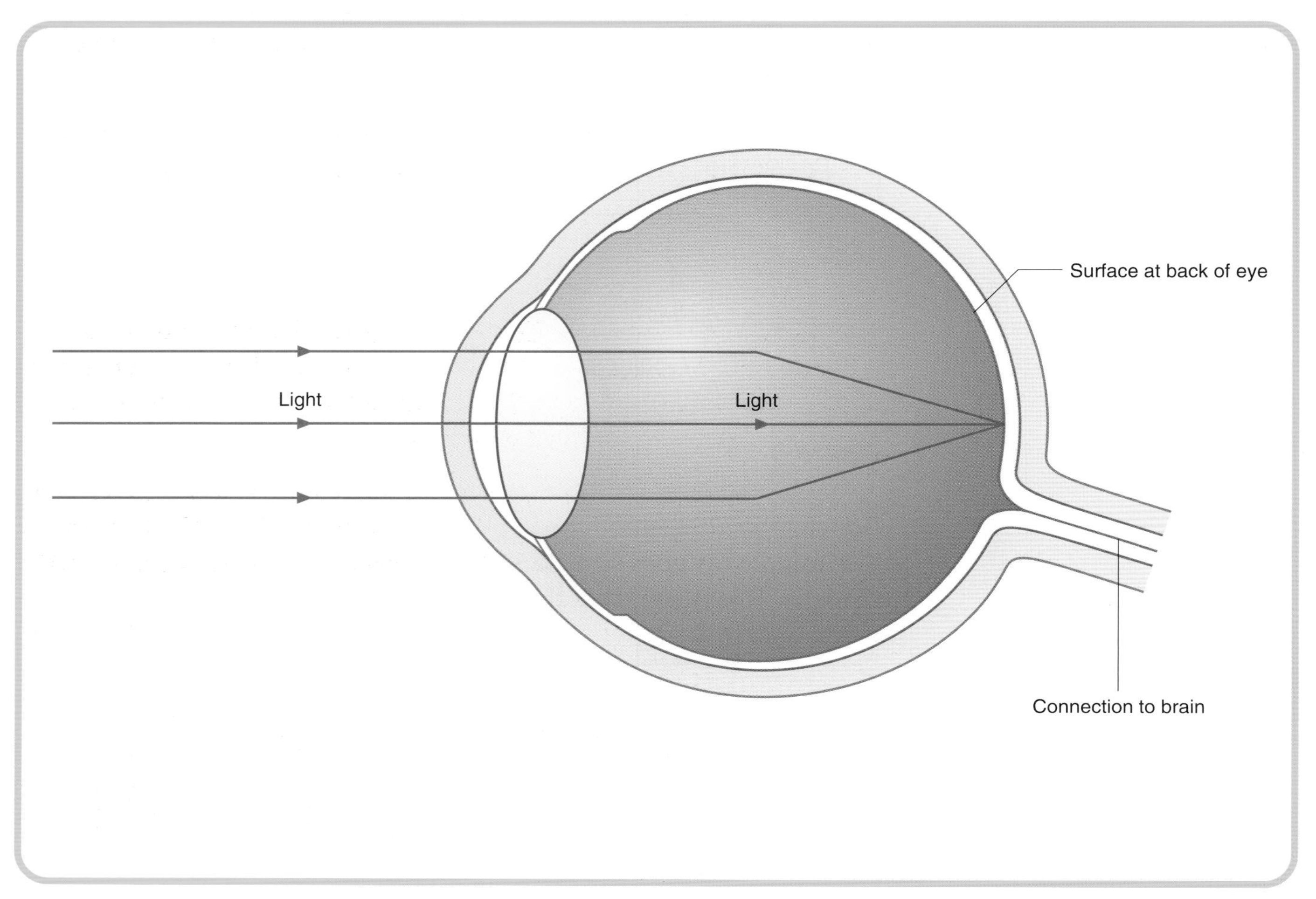


**Figure 3.16**

- Light enters the front of the eye at the cornea. This is transparent and most of the bending of light occurs here. The eye can detect changes from bright daylight to a dark night.
- The light enters a converging lens and more bending takes place. The lens is held in place by fibres which act like muscles. These can change the shape of the lens from thick to thin.
- The light then passes through a gel which makes the light spread out.
- The light reaches the retina at the back of the eye and this has special cells to receive the light.

Electrical signals pass along the nerve fibres to the brain. The part of the retina where the nerve fibres leave the retina contains no light-sensitive cells and is called a blind spot on the retina. To see clearly the light rays have to come to a focus on the retina.

To focus on near and on distant objects, the lens must change its shape. This is done by muscle-like fibres which can change the shape of the lens from thick to thin.

- To view distant objects clearly the lens must be thin.
- To view near objects clearly, the muscles change the lens shape to thick.

When light enters the eye, the image formed on the retina is upside down. The brain learns to turn this image the 'right way up'.

## Long sight

A long-sighted person can see distant objects clearly, but objects quite near to the eye appear blurred. The eye lens is bringing the rays to a focus beyond the retina. This is caused by the eye muscles not being able to make the lens fat enough.

A converging (convex) lens corrects this fault since it will increase the bending of the light rays before they enter the eye lens. The light rays will be focused on the retina and the image is seen clearly (figure 3.17).

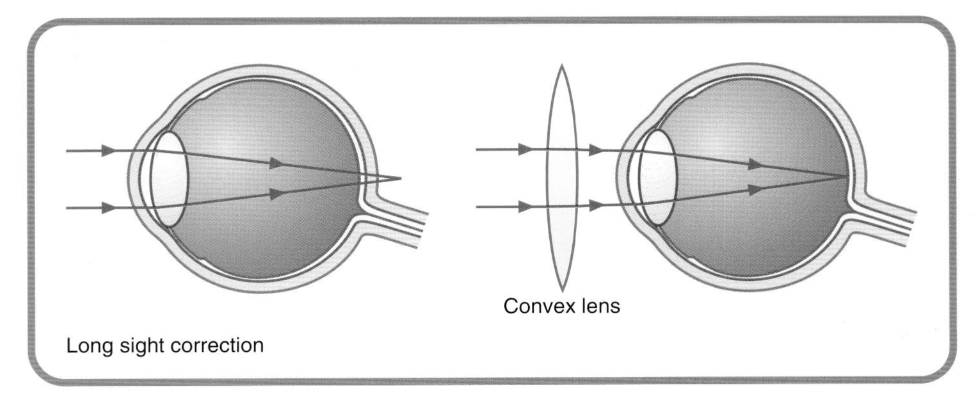


**Figure 3.17**

## Short sight

A short-sighted person finds that distant objects are blurred, but near objects are seen clearly. The eye lens is bringing light to a focus in front of the retina. This may happen due to the lens producing too much bending. The muscles cannot make the lens thin enough.

A diverging (concave) lens corrects this fault, since it will spread the light rays out more before they enter the eye lens. The light rays will be focused on the retina and the image will be seen clearly (figure 3.18).

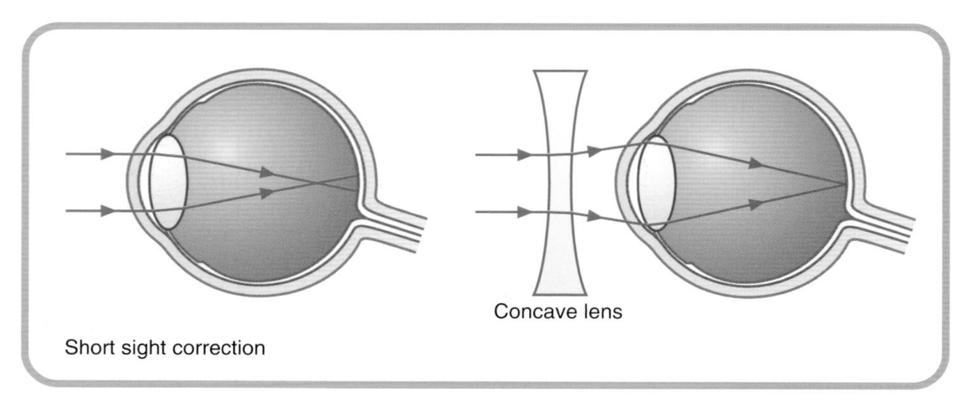


**Figure 3.18**

## Physics for you

### Contact lenses

Leonardo da Vinci first suggested contact lenses in 1581. He noticed that he could see more clearly when he opened his eyes under a bowl of water. Different lenses have been developed which allow oxygen to pass into the eye, which helps to prevent some eye diseases (figure 3.19).

These days contact lenses are very thin and can be made in different colours, which can give the eye a different appearance (figure 3.20).

Many people need to wear spectacles or contact lenses and we will nearly all need to have some help with seeing clearly like this as we grow older, but hopefully not just yet for you.

**Figure 3.19**

**Figure 3.20** *Coloured contact lenses*

Refraction is the term used when light passes into or out of a piece of glass. This is the bending of light. This effect can explain how some sight defects such as long and short sight occur and how they can be corrected.

## Key Points

- A laser is a concentrated source of light of only one colour.
- Lasers can be used as a scalpel or to correct eye problems.
- Light can be reflected.
- All visible objects give out, or reflect, light to the eye.
- A ray of light is reflected from the mirror at the same angle as it arrived at the mirror.
- Optical fibres are used in medicine to see inside the body.
- Converging lenses have a shape which bulges outwards in the middle and diverging lenses cave in.
- Converging lenses bring parallel light to a point called a focus and diverging lenses spread out parallel rays of light.
- In long sight, distant objects are seen clearly and near objects are blurred.
- In short sight, distant objects are blurred and near objects can be seen clearly.
- A converging lens can correct long sight and a diverging lens can correct short sight.

## Section Questions

**1** A torch has a coloured glass placed in front of it.
a) Is this a laser or just coloured light?
b) Explain your answer.

**2** State one use of lasers in medicine.

**3** Describe what happens when light reaches a mirror by copying and completing figure 3.21.

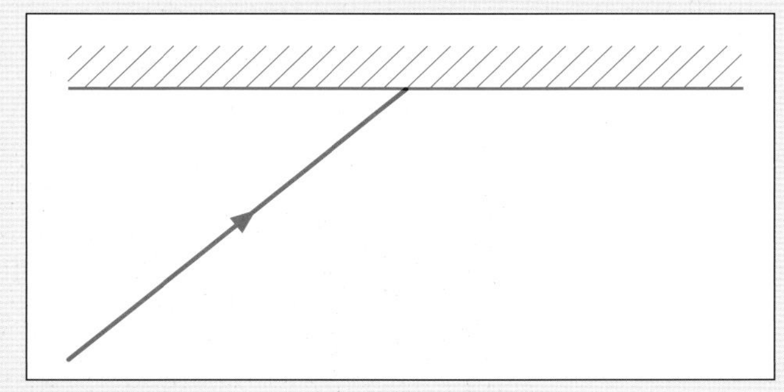

**Figure 3.21**

**4** Describe one use of optical fibres in medicine.

**5** Why is using fibre optics in medicine an advantage compared to other methods?

**6** a) A lens shape is shown in figure 3.22. What is this lens shape called?
b) Copy and complete figure 3.22 showing what happens to the rays of light after they pass through the lens.

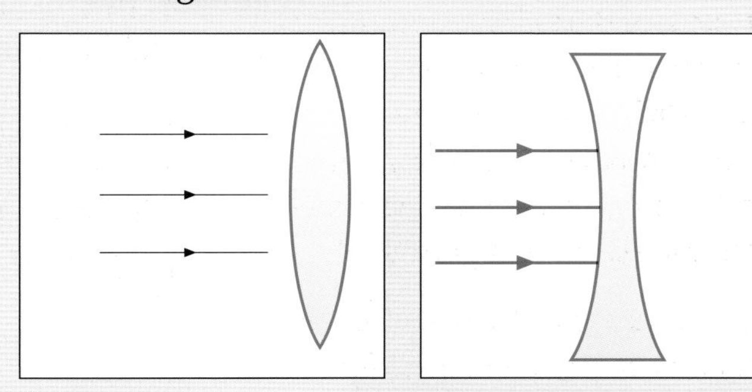

**Figure 3.22** **Figure 3.23**

**7** A diverging lens is shown in figure 3.23. What happens to the rays after they pass through the lens?

**8** Donna can see near objects clearly but distant objects are blurred.
a) State the sight defect she has.
b) What kind of lens is needed to correct this problem?

**9** Explain what is meant by a person being long sighted.

# 2 X-rays

## After studying this section you should be able to...

**1** State that X-rays are invisible to the naked eye.
**2** State that photographic film can be used to detect X-rays.
**3** Describe one use of X-rays in medicine.
**4** Describe one use of X-rays in industry.
**5** State that X-rays are dangerous since they can damage living cells.

### Medical X-rays

X-rays are used to either see inside the body or treat certain illnesses. They are produced in certain machines. We cannot see X-rays since they are not able to be detected by our eyes.

They are detected by photographic film like that used in ordinary cameras, but instead of colour film only black and white is used, since that is all that is needed in most types of medicine.

The use of X-rays in medicine depends on the fact that they pass through body tissues like skin, fat and muscle fairly easily, but are more easily absorbed by bones. When X-rays hit the photographic plate on the other side of the patient, they affect the photographic plate and blacken it. The image of an arm would be fairly dark, with lighter areas for the bone. The bones are lighter because they absorb the X-rays. The amount of blackening on the plate will depend on the number of rays reaching it.

Any break in a bone lets X-rays through and so may show up as a dark crack (figure 3.24).

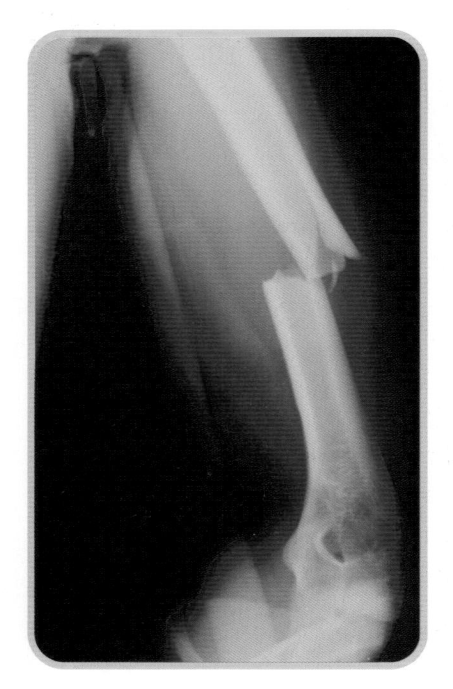

**Figure 3.24** *X-ray of a break in a bone.*

Some organs are made of the same kind of tissue and differences will not be clear. A substance like barium is used which absorbs X-rays. Barium is drunk by the patient to outline the stomach. The organs in the digestive system will show up lighter on the photographic plate (figure 3.25). Some X-ray machines do not use film to record results but instead use special detectors. The image produced is stored on a computer and displayed on a screen (figure 3.26).

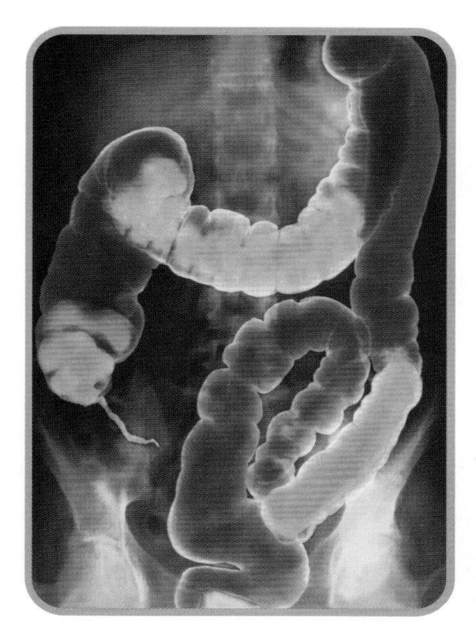

**Figure 3.25** *The digestive system.*

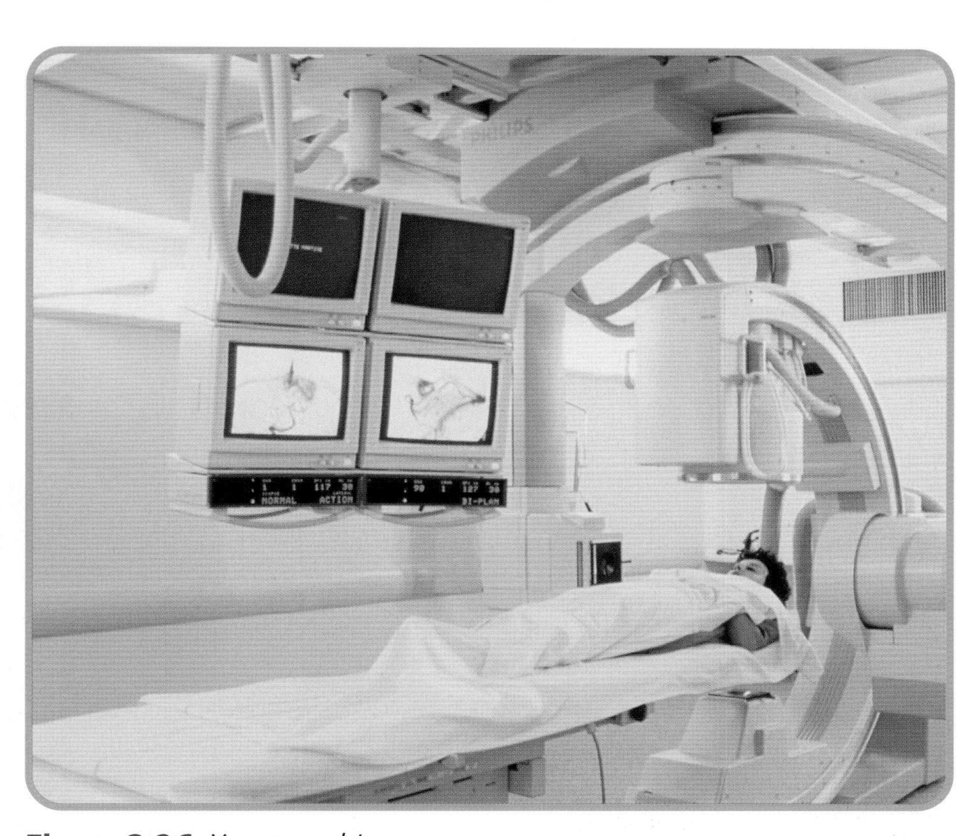


**Figure 3.26** *X-ray machine.*

X-rays are dangerous since they damage living cells. Normally we do not receive a lot of X-rays since we rarely come into contact with them, except in a normal X-ray following an injury. However if you work in hospitals as a radiographer then you could be exposed to large amounts of radiation. To prevent this radiographers are protected by wearing special aprons which are lead lined. These look like ordinary aprons but if you lift them you will find them to be very heavy. The lead absorbs the X-rays and prevents damage to the person. In addition the radiographers have special badges which can measure the amounts of radiation they receive over a time period (figure 3.27).

**Figure 3.27** *A badge that measures radiation.*

X-rays in industry are used to test any leaks that are very difficult to detect and costly to repair.

Since X-rays are absorbed by very heavy or thick objects they can be used to check for faults in equipment. If a water boiler is made for a ship, then it cannot be tested until it is put into the ship. An X-ray can be taken of any seals that were made in the boiler to test it before it is put into the ship (figure 3.28).

**Figure 3.28** *An x-ray machine being used to test the components of a ship.*

Another use is scanners at airports, which check your luggage before it is put on the aircraft. The image on the screen can show up anything with metal. A problem with this is that any photographic film in the cases will receive some X-rays and will be blackened slightly (figure 3.29).

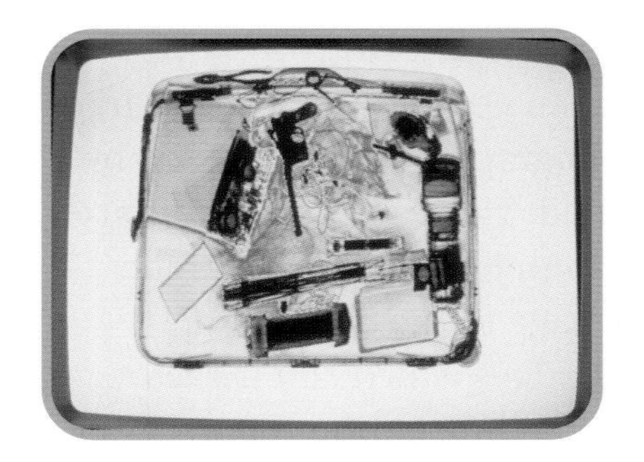

**Figure 3.29** *An airport scanner.*

## Key Points

- X-rays are invisible to the naked eye.
- Photographic film can be used to detect X-rays.
- X-rays are used in medicine to detect breaks called fractures in bones.
- X-rays are used in industry to detect flaws in metal structures which are invisible to normal eyesight.
- X-rays are dangerous since they can damage living cells.

## Section Questions

1. Fred breaks a leg and an X-ray is taken. He can see the machine but nothing else appears to be coming from the machine. What does this tell you about X-rays?
2. When an X-ray is taken, what is used to detect these rays?
3. Describe one use of X-rays in medicine.
4. A boiler is made to heat water for a ship. Before it is fitted it is checked for any faults in it. What is used to detect the faults?
5. Radiographers wear lead aprons when taking X-rays. Why do they need to wear this protection?

# Gamma rays

## After studying this section you should be able to...

1. State that gamma radiation is invisible to the naked eye.
2. State that gamma radiation can kill living cells or change the nature of living cells.
3. State that gamma radiation can pass through most materials.
4. Describe how gamma radiation can be used as a tracer in both medicine and industry.
5. State that the strength of a source of gamma radiation decreases with time.
6. Describe the safety precautions needed when dealing with a source of gamma radiation.
7. State that there is gamma radiation present in our surroundings.

**Figure 3.30** *A gamma radiation source.*

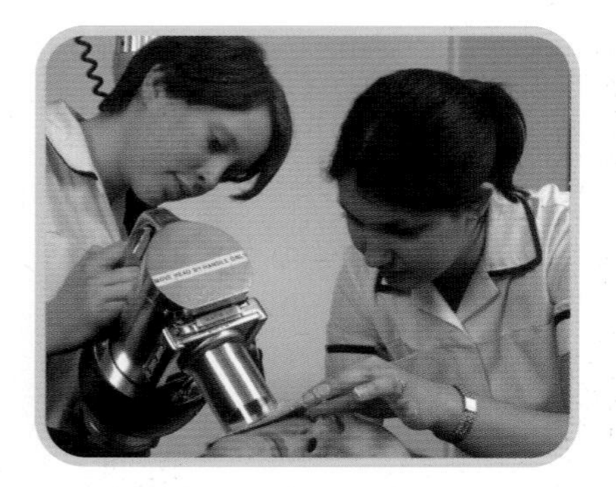

**Figure 3.31** *Gamma radiation being used on a patient.*

## Gamma radiation

Radiation is all around us but what exactly is radiation?

A source of gamma radiation is shown in figure 3.30.

Even if you look very closely you will not be able to see any radiation coming out of the source. If there was a very strong source in any building then you would not know unless you had a special machine which can be used to show radiation.

Gamma rays are invisible to our eyes but they can cause damage to your body.

You may be exposed to gamma radiation if you suffer from cancer. The radiation will try and kill the cancerous tumour. This will happen because the radiation will kill those cells which make up the tumour or else the radiation will change the cells so that they cannot spread the cancer around the body. This is done by putting the patient on a table and checking carefully that only the tumour is being exposed to radiation. The gamma radiation is then allowed to reach the patient for a short time. Some cells will now be destroyed. If the patient returns several times then each time the radiation will kill more of the cells until the tumour is destroyed (figure 3.31).

The strength of the radiation decreases as it passes through any material. Because it is so strong the radiation will pass through most materials. However, lead and concrete materials can absorb most of the radiation. This is why they are used to protect people.

## Tracing radiation

John had a problem with his kidneys. The doctor thought that there was a blockage or a tumour in one of them. If you had a blockage in a clear pipe you could easily see where it was by looking down the length of the pipe. If the liquid was cloudy then to see what was happening you could put a dye into the water and watch what happened. Where the dye stopped would be the source of the blockage.

A similar method can be used in medicine. Instead of operating to find any blockage, a radioactive substance which emits gamma radiation can be put into the body. The radioactive substance is mixed with a drug which is taken in by certain parts of the body, for example by the kidneys (figure 3.32). The way that the drug travels through the body can be watched by detecting the radiation emitted outside the body. This is done using a special camera which can detect the gamma radiation. This is called a gamma camera (figure 3.33).

Although there is a small amount of radiation in the body, it leaves the body very quickly and does no lasting damage.

**Figure 3.32**

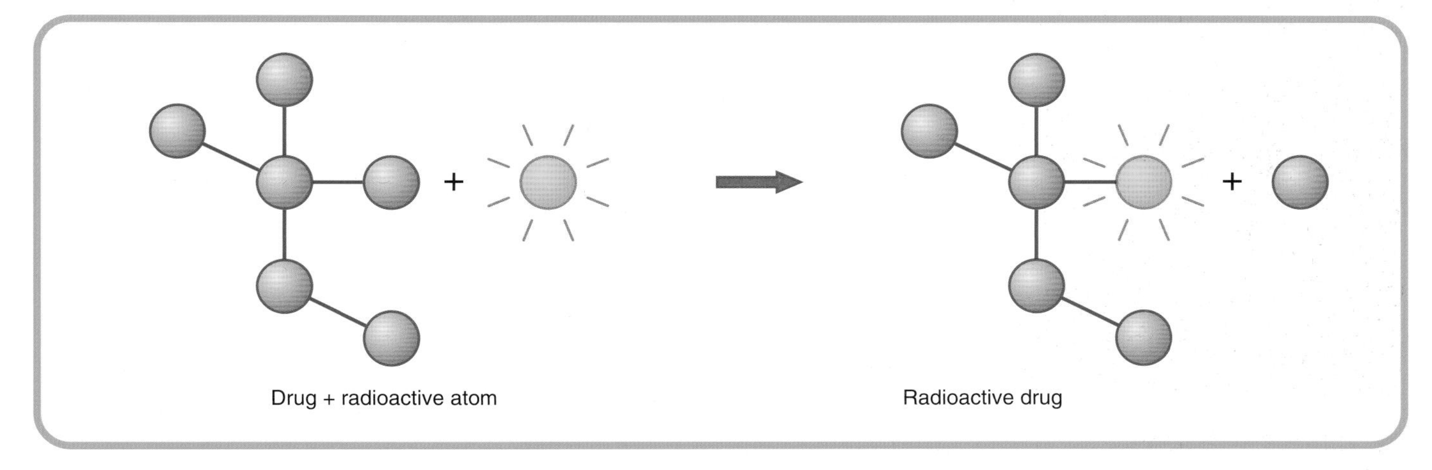


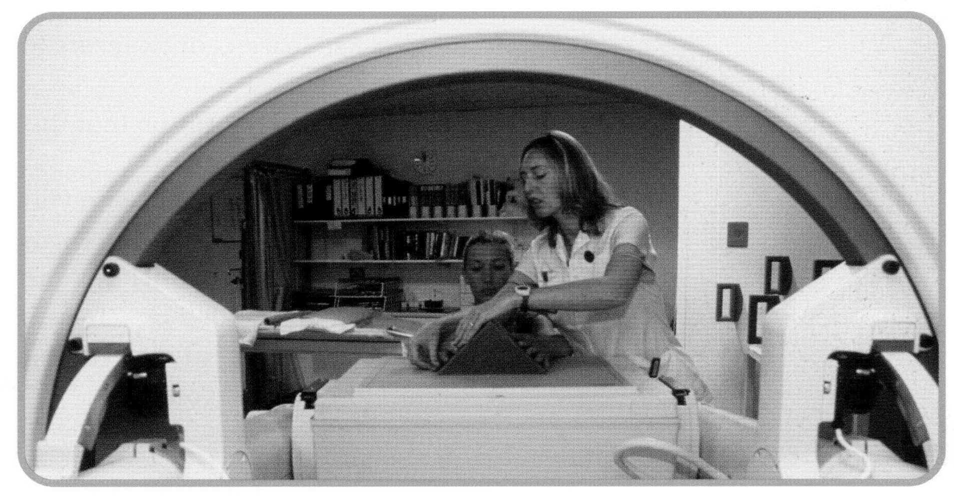

**Figure 3.33** *A gamma camera.*

**Figure 3.34** *Radioactive waste.*

## Time and gamma radiation

Gamma radiation will lose its strength over a period of time. This time can be a few seconds or millions of years. The strength of the radiation used as a tracer will decrease in a few hours. Some radiation used in treatments takes days for its strength to decrease. Radiation from the waste of nuclear reactors will take hundreds of years for its strength to reduce to a safe level (figure 3.34).

## Safety and gamma radiation

While radiation is dangerous, particularly from very strong sources, safety can help to reduce any dangers to most people. When radiation sources are being used or stored then a special sign tells us that they are in use, as shown in figure 3.35.

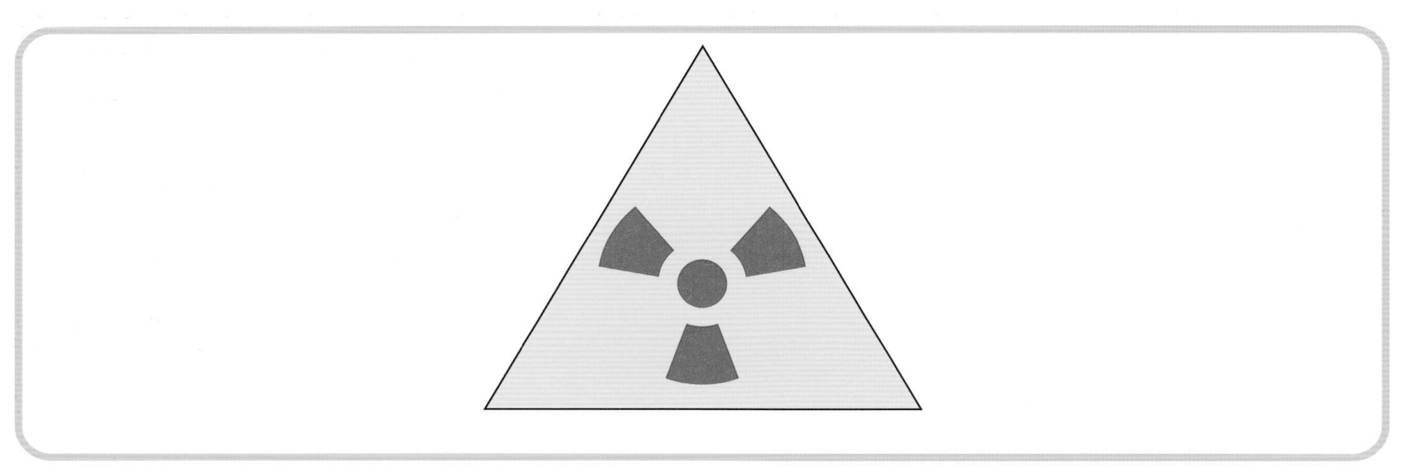

**Figure 3.35** *Sign for radioctive waste.*

## Safety with radioactivity

- Always use forceps or tweezers to remove a source. Never use bare hands.
- Always point a source so that the radiation  points away from the body.
- Never bring a source close to your eyes.
- When in use, a source must be used by someone who is trained in using it. The source must be returned to a locked and labelled store in its special shielded box immediately after use.
- After any experiment with radioactive materials, wash your hands thoroughly before you eat.
- In the UK, students under 16 years old may not normally handle radioactive sources.

## Key Points

- Gamma radiation is invisible to the naked eye.
- Gamma radiation can kill living cells or change the nature of living cells.
- Gamma radiation can pass through most materials.
- Gamma radiation can be used as a tracer in both medicine and industry by detecting the radiation as it passes along a body.
- The strength of a source of gamma radiation decreases with time.
- There are safety precautions needed when dealing with a source of gamma radiation.
- There is gamma radiation present in our surroundings.

## Section Questions

1 When a pupil visits a hospital he sees a source marked gamma radiation. Nothing appears to be coming from it yet it is dangerous. What does this tell you about gamma rays?

2 Gamma rays are used to treat cancerous tumours. What do the rays do to the cells of the tumour?

3 Gamma sources are kept in a lead box. Why are they not kept in an aluminium box which is lighter?

4 Sheila has a possible blockage affecting the blood flow in a kidney. Gamma radiation is mixed with a drug and injected into her. How is the flow of the drug checked as it flows along the body?

5 The strength of a source of radiation was measured in 1994 and again in 2004. What would you notice about the change in strength?

6 Describe two safety precautions you should take with radioactive sources.

7 A radiation detector is used to measure radiation at various parts of Scotland. There is a reading everywhere in Scotland. What does this tell you about gamma radiation?

# Infrared and ultraviolet

## After studying this section you should be able to...

**1** State that infrared radiation is invisible to the naked eye.
**2** State that infrared radiation is called heat radiation.
**3** Describe one use of infrared radiation in medicine.
**4** Describe one non-medical use of infrared radiation.
**5** State that ultraviolet radiation is invisible to the naked eye.
**6** Describe one use of ultraviolet radiation in medicine.
**7** State that some chemicals glow, that is fluoresce, when they absorb ultraviolet radiation.
**8** Describe how ultraviolet radiation can be used in identifying security markings.
**9** State that excessive exposure to ultraviolet radiation may produce skin cancer.

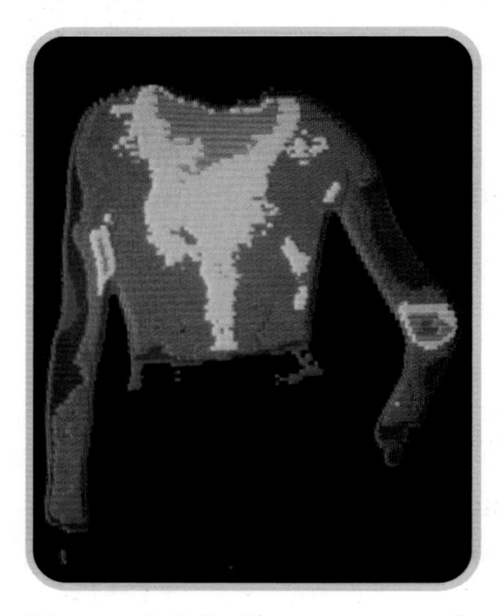

**Figure 3.36** *Thermogram of a body.*

**Figure 3.37**

## Infrared radiation in medicine

All hot objects give off invisible 'heat rays' called infrared radiation. Special cameras can be used to take colour photographs called thermograms using this radiation instead of light. The infrared radiation allows us to measure small temperature changes inside the body without surgery.

In medicine, thermograms of a patient's body show areas of different temperature. Doctors have found that cancerous tumours are warmer than healthy tissue and show up clearly on thermograms. This can be seen as different colours.

Some people suffer from arthritis which can make walking difficult, and a thermogram can show if certain drugs are helping. If the knee joint is affected then it will be slightly warmer than the other normal joint. A thermogram will show any temperature changes before and after using these drugs. Since blood changes temperature as it flows around the body then a thermogram can be used to look at blood flow in the body (figure 3.36).

Another use is the heat seeking cameras used to detect people who may be trapped in buildings. This saves the rescue teams time since they know where people are trapped and can dig at that place rather than in empty buildings (figure 3.37).

Infrared radiation is used in a different way by physiotherapists to treat people who have suffered a muscle injury. This is often used in sports injuries. A heat lamp is used to send this energy through the skin and heat the muscles and tissues. This heat causes healing to occur more quickly since the blood flow to the tissue is increased (figure 3.38).

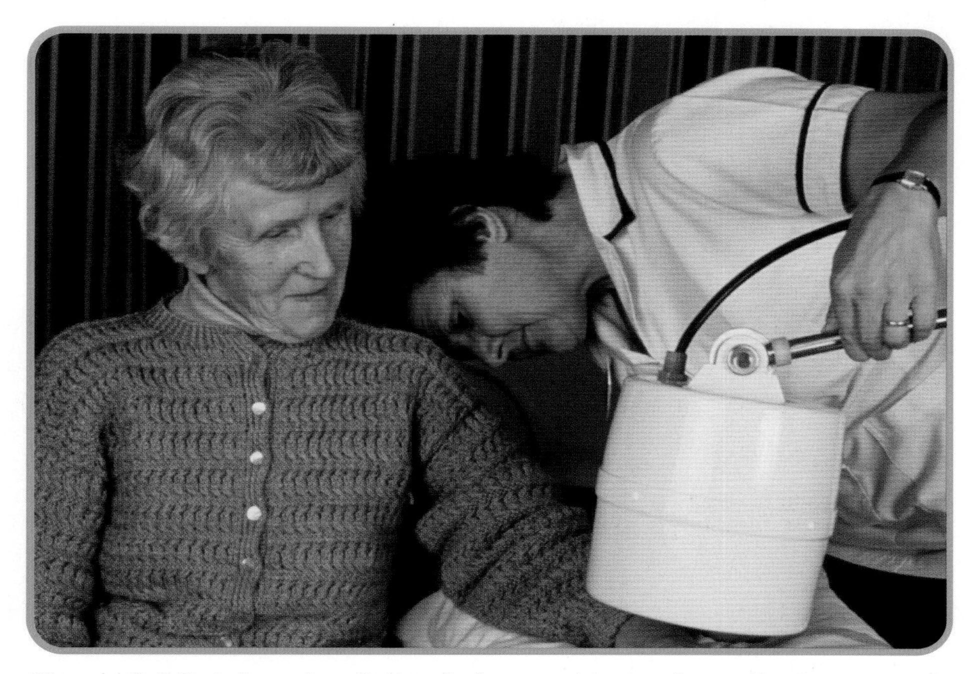

**Figure 3.38** *Infrared radiation being used to treat a patient.*

## Ultraviolet radiation

Ultraviolet (UV) is another type of invisible radiation.

Medicine uses this radiation to treat some kinds of skin diseases such as acne. With certain drugs and the UV radiation, the skin problems are reduced. Some sun beds can help these problems but if you use them for a long time they can cause problems.

Figure 3.39

## Glowing in the dark

Adverts often claim that certain washing powders are better at making white clothes appear to be whiter than white. This is because within the soap powder are chemicals which make the clothes glow or fluoresce when UV radiation is used. You can see this effect at discos when the lights make shirts glow with different colours (figure 3.39 ).

Fluorescent lamps can produce a bluish tinge in clothes. This is due to UV radiation.

Some shops will use a UV light source to check for forged banknotes. Certain marks will not be visible in forgeries but glow when exposed to UV radiation (figure 3.40).

Figure 3.40

You can use certain pens at home to security mark special items in case they are stolen. If you put any writing on the object it will only be visible under a UV source. This can be used to identify that the object is yours (figure 3.41).

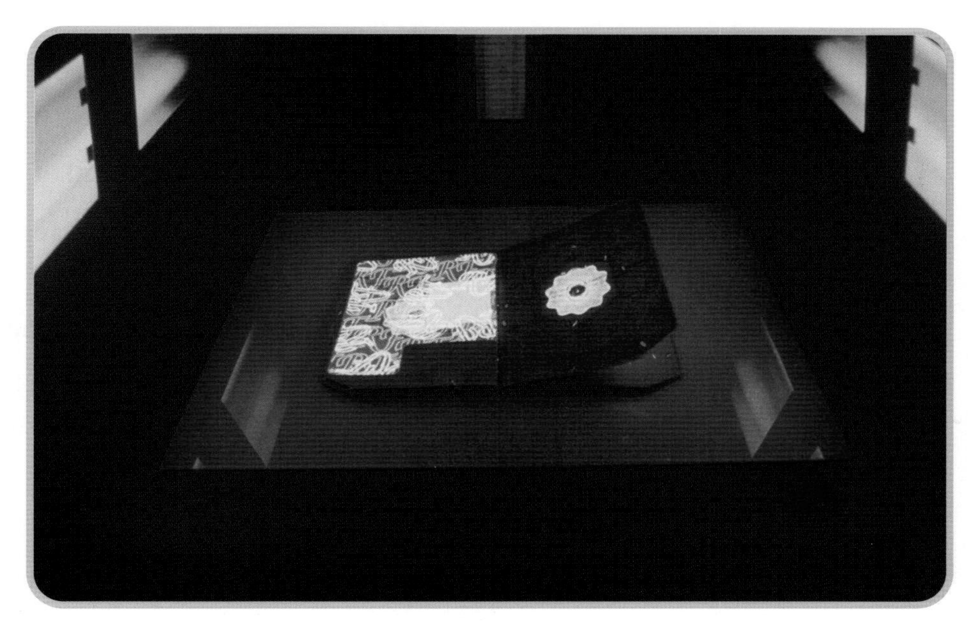

**Figure 3.41** *Security markings on a passport, showing up under UV light.*

Although ultraviolet forms only 3% of sunlight, it is the most damaging part. There are three types of UV radiation:

UVA, UVB and UVC.

All UVC and part of UVB are absorbed by the Earth's ozone layer. This is at the edge of the layers of air that form the atmosphere. UVB is absorbed by glass and many types of plastic. UVB is the part of the rays that cause sunburn and is more damaging than UVA.

We need UVA for healthy growth and to make vitamin D in our bodies.

UVA is not screened by glass and until recently suncreams were not effective against this type of radiation. UVA makes up more than 90% of the UV radiation that reaches the Earth's surface. A lot of exposure to UVA and UVB can cause skin cancers. There is also a risk from long exposure under sun beds and in this case care should be taken to cover the eyes since they are especially sensitive to this damage. Too much ultraviolet light on the skin produces sunburn and can cause the skin to turn red and be very painful.

UVA ages the skin and UVB burns the skin. This UV radiation is not felt as heat on the skin, so even on a cool and cloudy day you can still have UV radiation which can be dangerous.

This has been described as the ABC of skin damage:

**A**ged skin which looks ugly
**B**urnt skin – very painful
**C**ancer – can be fatal.

In the last few years the greatest rise in cancers in the UK has been as skin cancers.

Your skin can 'remember' the last amount of exposure to the sunlight that it received and over a period of time this may lead to skin cancer.

In the UK there are 40 000 new cases every year and it is the second most common cancer.

6000 of these will be the serious kind and 2000 people will die each year.

It is more common in people with a light coloured skin and who have been exposed to a lot of time in the sunlight. You will receive more sunlight if you are:

- sunbathing round about mid day,
- have fair or red hair,
- tan with difficulty and burn in the sun.

You can reduce the risk by:

- Limiting your time in the sun and covering up with loose clothing.
- Using a sunscreen lotion with a factor of 15 or greater. The number indicates the length of time that you can stay in the sun and seeking shade at the hottest part of the day.
- Checking the UV index given in weather forecasts.

A typical chart is shown in table 3.2.

| UV index number | Exposure level |
|---|---|
| 0 to 2 | Minimal |
| 3 to 4 | Low |
| 5 to 6 | Moderate |
| 7 to 9 | High |
| 10 + | Very high |

**Table 3.2**

Excessive UV radiation can also increase the chance of cataracts. These are changes to the lens of the eye which do not let light through the eye. The lens must be replaced with a plastic one. Suntan lotions absorb some of the ultraviolet rays which cause the burning, but they allow other rays to reach the skin and to produce a tan. The tan is due to a pigment called melanin being produced.

## Key Points

- Infrared radiation is invisible to the naked eye and is called heat radiation.
- One use of infrared radiation in medicine is to help injured muscles heal.
- A non-medical use of infrared radiation is to detect body heat from people trapped in buildings.
- Ultraviolet radiation is invisible to the naked eye.
- A use of ultraviolet radiation in medicine is to cure certain skin diseases.
- Some chemicals glow, that is fluoresce, when they absorb ultraviolet radiation.
- Ultraviolet radiation can be used in identifying security markings on equipment or banknotes.
- Excessive exposure to ultraviolet radiation may produce skin cancer.

## Section Questions

**1** To check on some diseases, the temperature of the organ is measured by measuring the heat produced by the organ. What is the name for this heat energy?

**2** Describe one use of this heat energy in medicine.

**3** During earthquakes, sometimes people are trapped in collapsed buildings. They can be detected by special cameras which detect invisible radiation given off by living people. Describe in a few sentences how these cameras can detect trapped people.

**4** Ultraviolet radiation is needed to keep us healthy. It can also be used in other ways in medicine. Describe one use of ultraviolet radiation.

**5** To protect computer equipment that might be stolen it can be marked with a special pen which makes invisible writing. When ultraviolet radiation is shone on the markings they can be seen. How is this possible?

**6** We all need some ultraviolet radiation, but when we are exposed to an excess of this radiation, sometimes we can develop a serious illness. What can happen if we are exposed to large amounts of ultraviolet radiation?

## Examination Style Questions

**1** Gamma radiation is used to check blood flow in an organ of the body.

a) In 5 minutes 3000 counts are recorded. Calculate the number of counts per minute.

b) The hospital checks the radioactive source again in 3 years. Will the number of counts per minute have decreased, increased or stayed the same after this time?

c) When the gamma source is not in the room, the radiation detector still measures a small number of counts per minute. Why does this happen?

d) The source is kept in a lead lined container but care has to be taken in using this source. State one precaution to be carried out when using such sources.

**2** A powerful light source is covered with a piece of coloured plastic. The source is placed 1 metre from a light meter. This measures the amount of light from a source (figure 3.42). A measurement is taken and the experiment repeated with a laser light. Results are shown table 3.3.

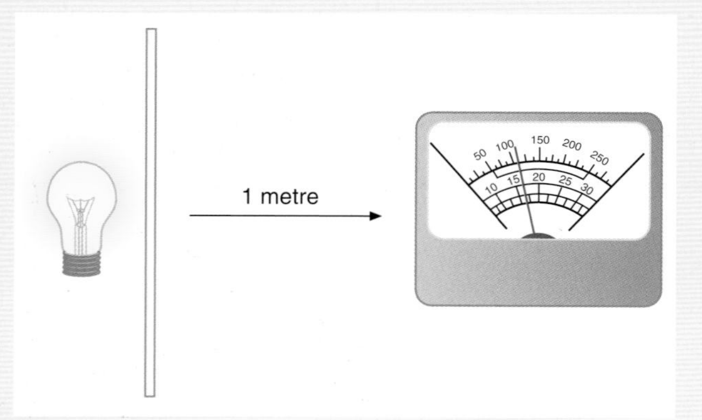


**Figure 3.42**

| Source | Reading in light units |
|---|---|
| A | 300 |
| B | 1750 |

**Table 3.3**

a) Which source is the laser source?

b) Explain your answer.

c) The light source is used in an endoscope. State one use of this in medicine.

d) The light energy is sent to a camera and has to be brought to a sharp focus by a lens. The camera uses a converging lens. Explain what happens to rays after passing through a converging lens.

**3** If you receive an injury at sport, you may receive treatment using infrared radiation.

a) How will your body detect this radiation?

b) On holiday, your luggage passes through a scanner. What kind of radiation is used to detect any problems with luggage?

c) To make sure you have a tan you decide to sunbathe for a short time each day. What kind of radiation produces a suntan?

d) If you receive more of this type of radiation than you need, what kind of danger might you be exposed to?

**4** a) Draw a diagram to show the shape of a converging lens.

b) Chris cannot see distant objects clearly but can see the writing on this page easily. What sight defect does he have?

c) What kind of lens would correct this sight defect?

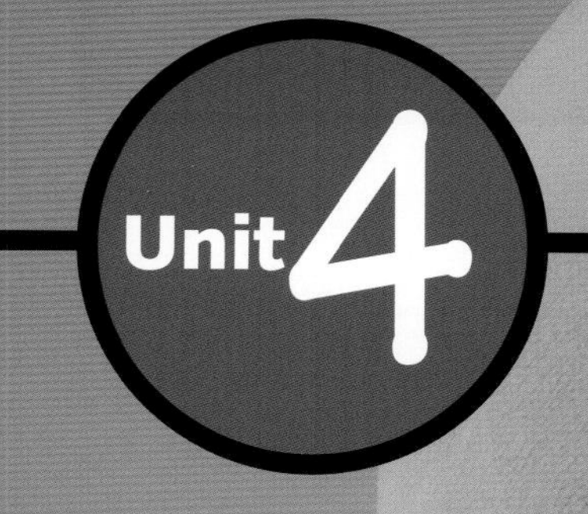


# Sound and Music

**Sound and music are important parts of our life, both for communication and entertainment. This chapter considers the sounds around us – how sounds are produced and how they can be altered. The difference between the speed of sound and the speed of light is discussed. The use of sound in medicine and industry and the effects of noise pollution are considered.**

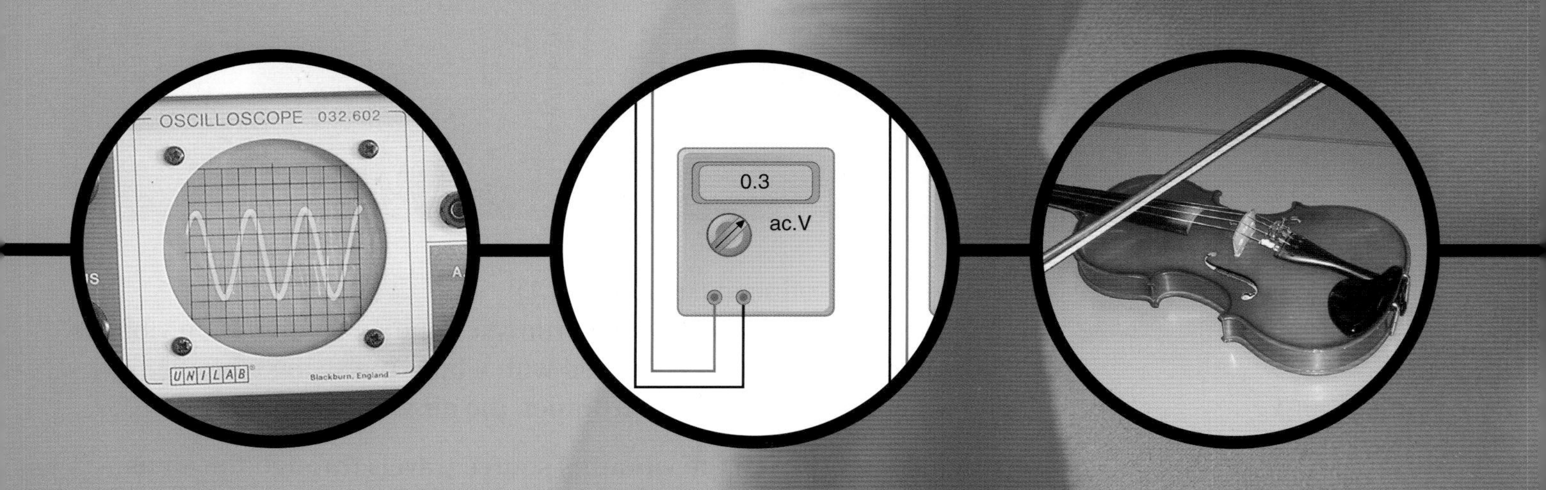

# 1 Sound waves

## After studying this section you should be able to...

1 State that for sound to be produced an object must vibrate.
2 State that sound is a wave which transfers energy.
3 State that the frequency of a sound is the number of waves produced in 1 second.
4 State that frequency is measured in hertz.
5 State that the higher the pitch of a sound the larger the frequency.
6 Identify from oscilloscope traces the signal which would produce:
  a) the louder sound,
  b) the higher frequency.
7 State that if two sounds are one octave apart, the frequency of one is double the other.
8 State that the frequency produced by a vibrating string can be increased by shortening the length of the string and increasing the tightness of the string.
9 State that the frequency produced by a vibrating air column can be increased by shortening the length of the air column.

We are surrounded by sounds — but what is a sound?

Sound is a form of energy, just like heat and light. A sound is produced when something vibrates — moves backwards and forwards or shakes.

**Figure 4.1**

A sound is produced (see figure 4.1):

- when a drum is struck with a drumstick, the skin of the drum vibrates,
- when a violin string is rubbed with a bow, the string vibrates,
- when air is blown into a trumpet, the air vibrates.

Sound energy from the vibrating source travels through the air as a wave.

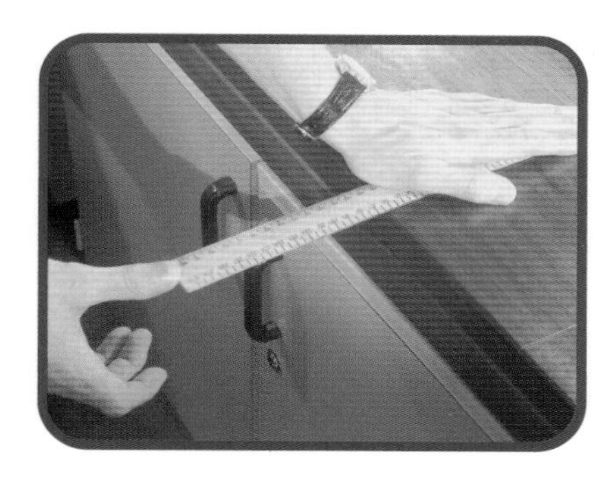

Figure 4.2

The faster the vibrations, the more waves are produced in 1 second. The number of waves in 1 second is called the frequency.

Clamp a ruler to a table using your hand (figure 4.2). Investigate how the frequency and loudness of a sound can be changed.

Objects which vibrate slowly produce low frequency sounds — a few waves per second.

Objects, which vibrate quickly, produce high frequency sounds — a lot of waves per second.

The larger the vibration the louder the sound.

Use a sonometer or a stringed instrument to investigate how the frequency and loudness of a sound can be changed (figure 4.3).

**Figure 4.3** *A sonometer.*

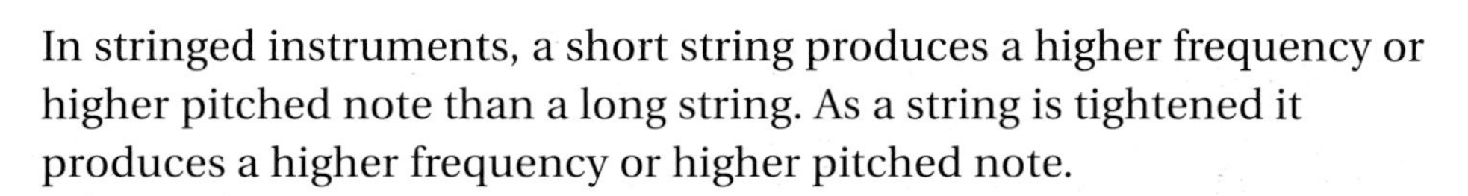

In stringed instruments, a short string produces a higher frequency or higher pitched note than a long string. As a string is tightened it produces a higher frequency or higher pitched note.

Figure 4.4

Fill a glass bottle with some water so that you have an air column inside the bottle.

Tap the bottle carefully with a metal object and note the sound produced (figure 4.4). Add more water to the bottle so that the air column is smaller and tap the bottle.

Continue adding water to the bottle and tapping to find how the length of air column changes the frequency of the sound produced.

In wind instruments, such as a whistle, recorder or a trumpet, a short column of vibrating air produces a higher frequency or higher pitched note than a long column of air.

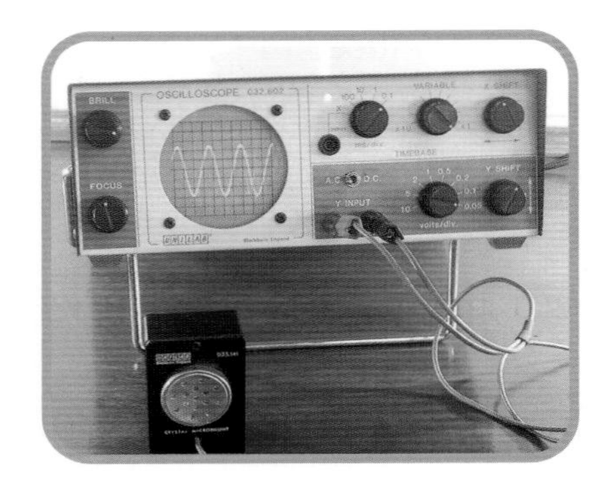

**Figure 4.5** *An oscilloscope connected to a microphone.*

## 'Seeing' sound waves on an oscilloscope

Connect a microphone to an oscilloscope (figure 4.5).

Whistle different frequencies into the microphone. Look at the different traces on the oscilloscope screen.

Now whistle loud and quiet sounds into the microphone. Look at the different traces on the oscilloscope screen.

The traces on an oscilloscope screen obtained from two sounds of the same loudness but different frequency are shown in figure 4.6.

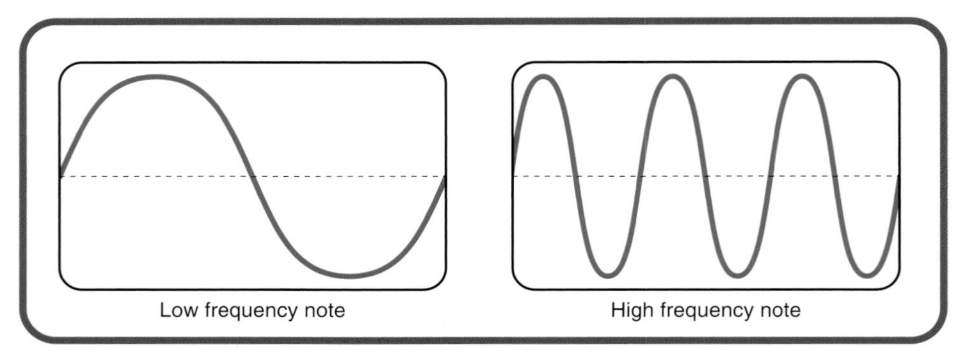


**Figure 4.6**

When the frequency of a sound changes, the number of waves on an oscilloscope screen changes.

The higher the frequency the more waves are seen on the screen of the oscilloscope.

The traces on an oscilloscope screen obtained from two sounds with the same frequency (pitch) but different loudness are shown in figure 4.7.

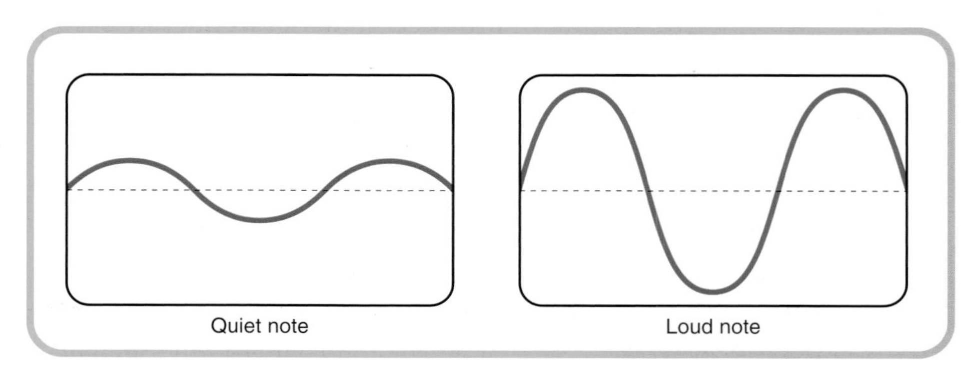


**Figure 4.7**

When the height of a trace on an oscilloscope screen changes, the loudness changes.

The louder the sound, the bigger the trace on the oscilloscope screen.

## Octaves

The note called middle 'C' on a piano has a frequency of 256 hertz. Upper 'C' on a piano has a frequency of 512 hertz. This is one octave above middle 'C'. Upper 'C' has twice the frequency of middle 'C'.

When two notes are one octave apart the frequency of the higher note is twice the frequency of the lower one.

### Examples

One string on a cello produces a note of frequency 1024 hertz.
a) How many sound waves does the string produce in 1 second?
b) Another string on the cello produces a note one octave higher. What is the frequency of this note?
c) A third string produces a note two octaves above the first string. What is the frequency of the note produced by the third string?

Solution
a) Frequency = number of waves in 1 second, so 1024 waves are produced in 1 second.
b) One octave higher = 2 x 1024 = 2048 hertz.
c) One octave higher = 2048 hertz,
Two octaves higher = 2 x 2048 = 4096 hertz.

### Key Points

- Frequency of a sound = number of waves in 1 second.
- Frequency is measured in hertz.
- Sound signal on an oscilloscope screen: (a) the more waves shown on the screen the higher the frequency, (b) the bigger the trace shown on the screen the louder the sound.
- The frequency of a sound is the same as the pitch – a high pitched sound has a high frequency.
- A short string produces more waves per second than a long string. A short string produces a note of higher frequency (pitch) than a long string.
- Increasing the tightness of a string produces more waves per second and so a higher frequency is produced.
- The shorter the length of air in a wind instrument the more waves are produced per second and the higher the frequency produced.
- When two notes are one octave apart then the frequency of the higher note is twice the frequency of the lower one.

## Section Questions

**1** Copy and complete the following sentences using appropriate words from the list: frequency, hertz, one, vibrates, wave.

When an object ______ a sound wave is produced. Energy from the vibrating object travels out as a sound ______. The number of waves produced in ______ second by the vibrating object is called the ______. Frequency is measured in units called ______.

**2** Copy and complete the following sentences using appropriate words from the list:
lower octave pitch twice.

The frequency of a sound is often referred to as the ______ of a sound. When two notes from a musical instrument are one ______ apart then the higher pitched note has ______ the frequency of the ______ pitched note.

**3**

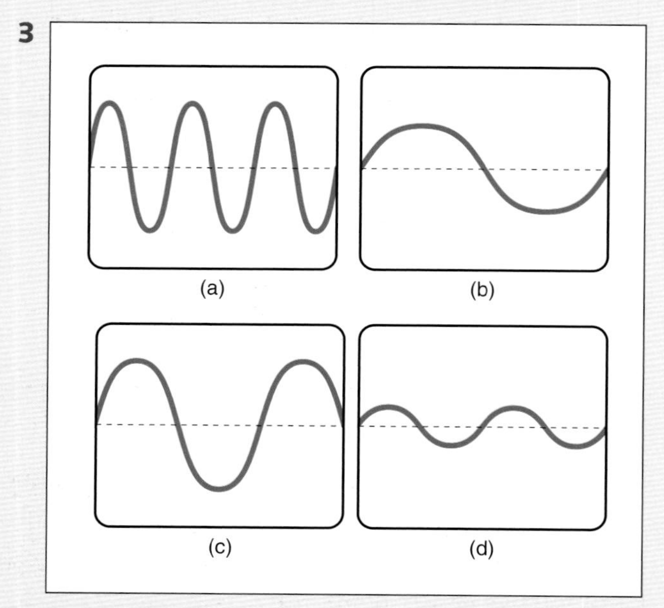


**Figure 4.8**

A microphone is connected to an oscilloscope. Different traces obtained from the oscilloscope are shown in figure 4.8. Which diagram:

a) Shows the loudest note?
b) Shows the quietest note?
c) Shows the note with the highest frequency?
d) Shows the note with the lowest frequency?

**4** The frequency of a note is 440 hertz. What is the frequency of a note:
a) One octave higher?
b) One octave lower?

**5** What is the frequency of a note two octaves lower than 1760 hertz?

**6** A violin is being tuned. What will happen to the frequency of the note produced when the string being played is:
a) Increased in length?
b) Tightened?

**7**

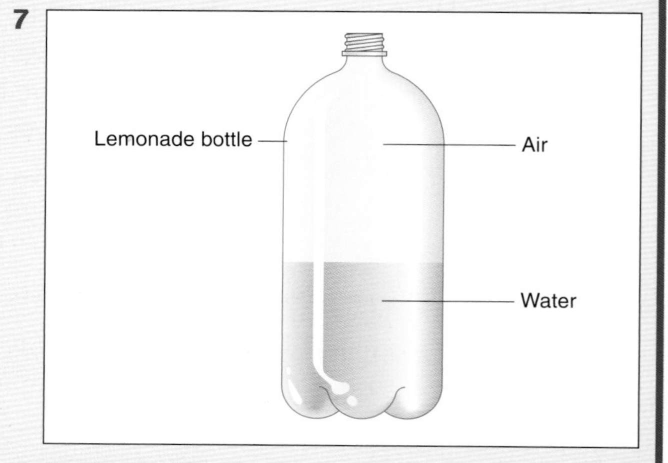


**Figure 4.9**

A girl constructs a simple musical instrument. She half fills a glass lemonade bottle with water and then taps it with a metal spoon (figure 4.9). She hears a certain frequency of sound from her musical instrument.

The girl now adds more water to the bottle and taps it again. How does the frequency of the new sound compare with the frequency of the original sound?

**8** A microphone is connected to an oscilloscope. A sound wave is played into the microphone and the trace obtained on the oscilloscope screen is shown in figure 4.10(a). The controls on the oscilloscope are not changed. The sound is changed and the trace on the oscilloscope screen now appears as in figure 4.10(b).

What differences are there between the two sound waves played into the microphone?

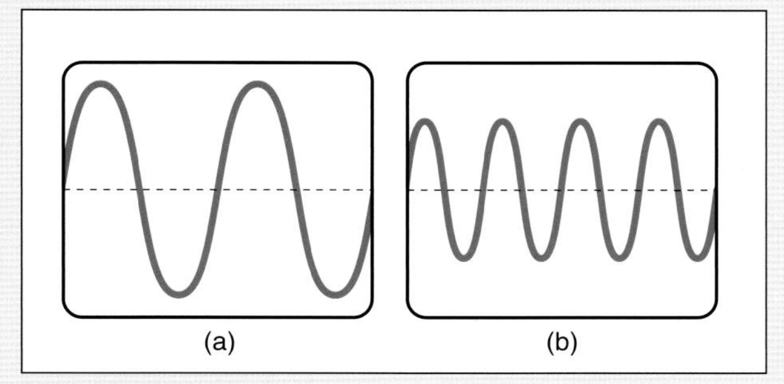


**Figure 4.10**

**9** John connects a microphone to an oscilloscope to look at the waveforms obtained from a musical tape. He observes three different traces on the oscilloscope screen as shown in figure 4.11.

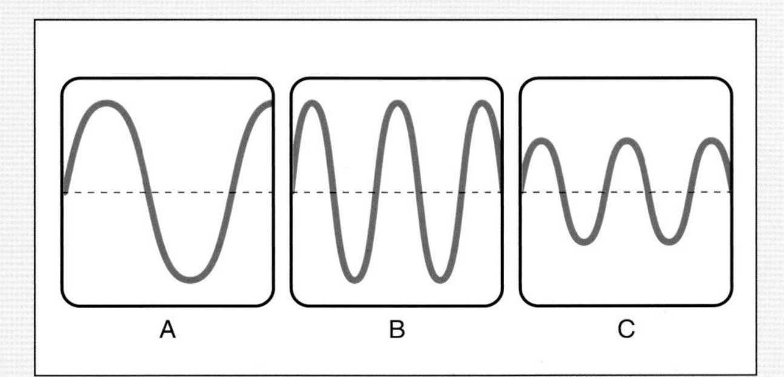


**Figure 4.11**

a) Which note, A or B, has the higher pitch?
b) Which note is louder, B or C?

**10** The unit of frequency is:
A second
B metre
C ampere
D hertz
E volts.

**11** The frequency of vibration of a guitar string is too low. The frequency of vibration can be made higher by:
A using a longer string
B tightening the string
C plucking the string harder
D plucking the string more softly
E using a thicker string.

**12**

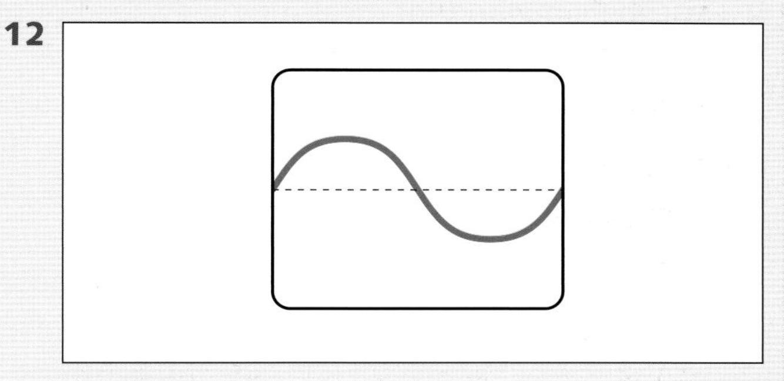

**Figure 4.12**

A microphone is connected to an oscilloscope. A sound of frequency 600 hertz is played into the microphone. The trace on the oscilloscope screen is shown in figure 4.12.

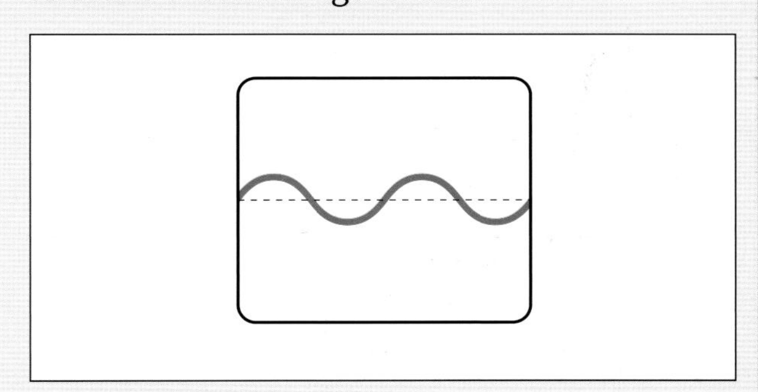

**Figure 4.13**

A sound of unknown frequency is now played into the microphone. The controls of the oscilloscope are not changed. The trace on the oscilloscope is shown in figure 4.13.

The frequency of the unknown signal is:
A 200 hertz
B 400 hertz
C 800 hertz
D 900 hertz
E 1200 hertz

# Speed of sound

## After studying this section you should be able to...

1. Give an example which shows that the speed of sound in air is less than the speed of light in air.
2. Describe a method for measuring the speed of sound in air using the relationship between distance, time and speed.
3. Calculate the speed of sound using: speed = $\frac{\text{distance}}{\text{time}}$.

During a thunderstorm, lightning and thunder are produced at the same time. However, when you are some distance from the storm, you see the lightning and then a short time later you hear the thunder.

From this we can say that the speed of sound in air is less than the speed of light in air.

But what is speed? Speed is the distance travelled by an object in 1 second.

### Examples

Fiona runs from A to B.

She covers a distance of 200 metres in a time of 40 seconds.

Fiona travels 200 metres in 40 seconds

Fiona travels $\frac{200}{40}$ metres in 1 second

Speed of Fiona = $\frac{200}{40}$ = 5 metres per second.

In the example 200 metres was the distance travelled and 40 seconds was the time taken:

speed = $\frac{\text{distance travelled}}{\text{time taken}}$.

### Examples

A sound wave takes 4 seconds to travel 6000 metres through water. Calculate the speed of sound in water.

Solution

speed = $\frac{\text{distance travelled}}{\text{time taken}}$.

= $\frac{6000}{4}$ = 1500 metres per second

## Measuring the speed of sound in air

Try to carry out the following three methods to measure the speed of sound in air. Think about any improvements you could make.

<u>Method A</u>

Apparatus needed: stopwatches, measuring tape.

1. A person stands at one end of a field with two pieces of wood. When the pieces of wood are clapped together a loud sound is produced.
2. At the other end of the field are some students with stopwatches.
3. When the timekeepers see the person clapping the pieces of wood together they start their stopwatches — this is the instant the sound is produced.
4. When the timekeepers hear the sound, the stopwatches are stopped and the times noted.
5. Measure the length of the field in metres using the measuring tape.
6. Calculate an average time for the sound to travel the length of the field.
7. Calculate the speed of sound in air using:

$$\text{speed of sound} = \frac{\text{distance travelled}}{\text{time taken}}.$$

Is this an easy experiment to carry out? What difficulties did you have?

<u>Method B</u>

Apparatus needed: stopwatches, measuring tape.

1. The timekeepers and the person who is going to produce the sound stand side by side some distance from a tall building (figure 4.14).

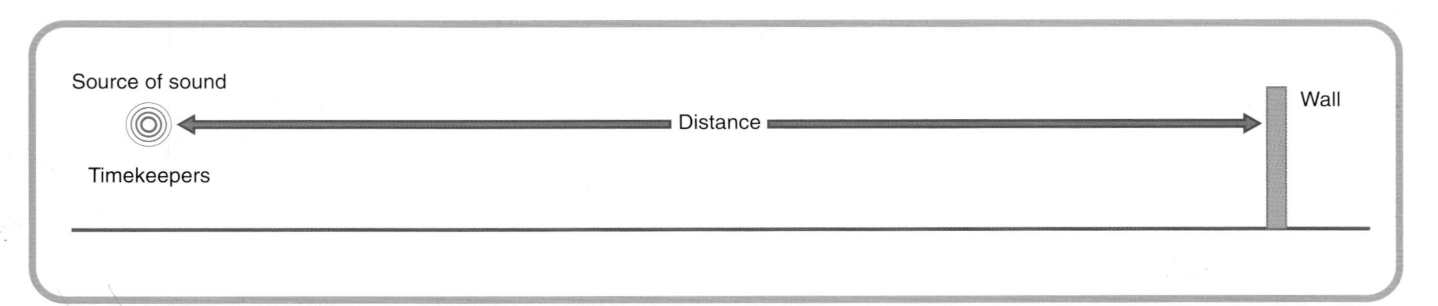


**Figure 4.14**

2. When the sound is produced the stopwatches are started. When the echo of the sound is heard, the stopwatches are stopped.
3. The distance between the timekeepers and the wall is measured in metres using the measuring tape. The distance travelled by the sound is the distance to the wall and back again. This is because the sound travels from the source of sound to the wall and the same distance back again to the timekeepers.

**4** Calculate the speed of sound in air using:

$$\text{speed of sound} = \frac{\text{distance travelled}}{\text{time taken}}.$$

**Note:** In the echo method the distance travelled is from you **to the wall and back again.**

What difficulties did you have in doing both methods A and B?

Why do you think these methods generally give an inaccurate measurement for the speed of sound in air?

Method C

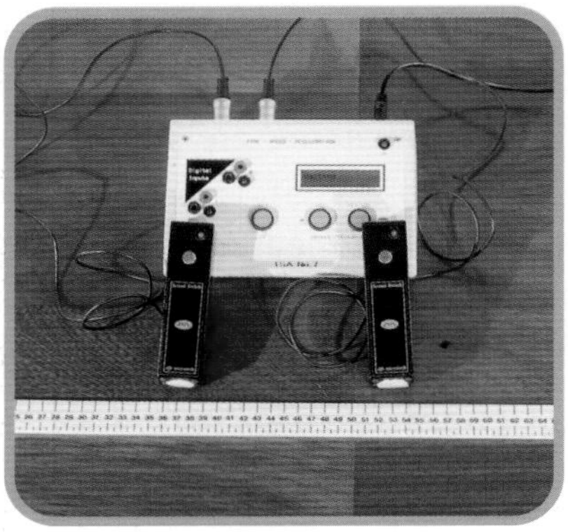

Apparatus needed: two microphones, electronic timer.

**Figure 4.15** *A timer and two microphones.*

1. Place two microphones 1 metre apart (figure 4.15).
2. Switch on the timer and make a sound in front of the first microphone When the sound reaches the first microphone, the timer starts timing. When the sound reaches the second microphone, the timer stops timing.
3. Note the time for the sound to travel from the first to the second to microphone.
4. Calculate the speed of sound in air using:

$$\text{speed of sound} = \frac{\text{distance travelled}}{\text{time taken}}.$$

The speed of sound in air has an approximate value of 340 metres per second.

Which of the three methods A, B or C was the most accurate?

## Key Points

- The speed of light is faster than the speed of sound. Light from lightning is seen before sound from thunder is heard.
- $\text{Speed of sound} = \frac{\text{distance travelled}}{\text{time taken}}.$

## Section Questions

1. You tell your cousin that the speed of sound in air is less than the speed of light in air. Give an example to support your statement.
2. Describe a method for measuring the speed of sound in air. Your answer should include
   a) a labelled diagram,
   b) a description of the measurements you would make,
   c) how you would use these measurements to find the speed of sound in air.
3. Write down the equation that lets you calculate speed.

**4** Copy and complete table 4.1.

| | Distance | Time | Speed |
|---|---|---|---|
| a) | 12 metres | 3 seconds | |
| b) | 100 metres | 5 seconds | |
| c) | 1500 metres | 300 seconds | |
| d) | 850 metres | 2.5 seconds | |

**Table 4.1**

**5** In an experiment to measure the speed of sound in air the following measurements were made.

Time taken for the sound to travel from the source to the timekeeper = 2 seconds.

Distance from source to timekeeper = 680 metres.

Use the information to obtain a value for the speed of sound in air.

**6** A student connects two sound operated switches to a timer as shown in figure 4.16.

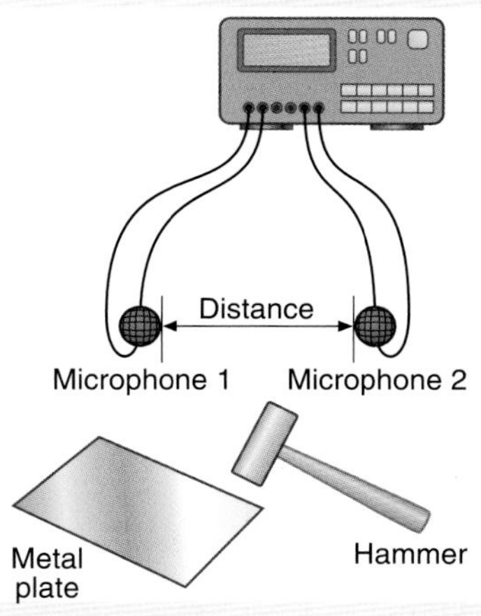


**Figure 4.16**

A sound is produced. When the sound wave reaches microphone 1, the timer starts timing. When the sound wave reaches microphone 2 the timer stops timing.

The student records the measurements shown in table 4.2.

| Distance between sound operated switches (metres) | 0.5 | 1.0 | 1.5 | 2.0 | 2.5 |
|---|---|---|---|---|---|
| Time for sound to travel between switches (seconds) | 0.0015 | 0.0030 | 0.0044 | 0.0059 | 0.0074 |

**Table 4.2**

a) Present the information in the table as a line graph.

b) Calculate the speed of sound in air when the time taken for the sound to travel between the microphones was 0.0059 seconds.

**7** A man standing 1700 metres away from a lightning strike hears the thunder 5 seconds after he sees the lightning. Calculate the speed of sound in air on this night.

**8** A hammer hits a metal pipe. The sound from the hammer blow takes 0.25 seconds to travel 1300 metres along the pipe. Calculate the speed of sound in the pipe.

**9** A girl stands 255 metres in front of a high, wide wall. She shouts and hears the echo 1.5 seconds later.

a) What is the distance travelled by the sound in 1.5 seconds.

b) Calculate the speed of sound on that day.

**10** Two students stand some distance from each other. One student makes a loud sound by clapping two pieces of wood together. The other student starts a stopwatch on seeing the pieces of wood coming together and stops the stopwatch on hearing the sound. To calculate the speed of sound, what other measurement must be made?

**11** a) A girl shouts into a canyon and a short time later she hears an echo. What causes this echo?

b) The girl moves to a different point and again shouts into the canyon. She now hears two echoes. What is happening this time?

**12** During a firework display the flash from a rocket exploding is seen and 0.5 seconds after this the sound is heard. The rocket exploded 170 metres from the crowd.

Calculate the speed of sound in air on the evening.

# 3 Using sound

## After studying this section you should be able to...

1. State that sound can pass through solids, liquids and gases.
2. State that sound cannot pass through a vacuum.
3. State that the normal range of human hearing is from 20 hertz to 20 000 hertz.
4. State that high frequency sounds beyond the range of human hearing are called ultrasounds.
5. Give one example of a use of ultrasound in medicine.
6. Give one example of a non-medical use of ultrasound.
7. State that sound levels are measured in decibels.
8. Give two examples of noise pollution.
9. State that excessive noise can damage hearing.

Design an experiment to show that a sound wave can travel through (a) a solid, (b) a liquid.

Design an experiment to show that a sound wave cannot travel through a vacuum.

Before sound can be produced an object must be made to vibrate. Solids, liquids and gases are made up of particles. A sound wave makes the particles vibrate and this allows the sound to pass through the solid, liquid or gas. There are no particles in a vacuum. Since there are no particles for the sound wave to vibrate, sound cannot pass through a vacuum.

## Examples

Table 4.3 shows the speed of sound through a number of different materials.

| Material | Speed of sound (metres per second) | Material | Speed of sound (metres per second) |
|---|---|---|---|
| Air | 340 | Water | 1500 |
| Blood | 1570 | Bone | 4100 |
| Steel | 6000 | Helium | 970 |
| Oil | 1400 | Carbon dioxide | 260 |
| Concrete | 3700 | | |

**Table 4.3**

Use the information in the table to answer the following questions.
a) In which type of substance does sound travel slowest?
b) In which type of substance does sound travel fastest?

Solution
a) Sound waves travel slowest in gases (slowest in carbon dioxide, air then helium).
b) Sound waves travel fastest in solids (fastest in steel, concrete then bone).

Sound waves travel fastest in solids and slowest in gases.

## The stethoscope

A stethoscope is used to listen to sounds made inside the human body.

Your teacher may have a stethoscope you could use to listen to sounds.

Place the bell of the stethoscope on your partner's back and listen to their heartbeat. Count the number of beats per minute using the stethoscope.

Place the bell of the stethoscope on the bench. Get your partner to scratch the bench lightly with a fingernail. How did the sound heard through the stethoscope compare with the sound heard without it?

A stethoscope makes the sounds made inside the body much louder and easier to hear. This can allow a doctor to identify certain illnesses.

## The range of human hearing

Your teacher may be able to test your upper limit of hearing and possibly your lower limit of hearing.

Connect a signal generator to a loudspeaker. Set the frequency control on the signal generator to about 25 000 hertz and switch on. Slowly lower the frequency control on the signal generator. When you can first hear the sound from the loudspeaker, note the frequency. This frequency is your upper limit of hearing.

Most young people can hear sounds from a frequency as low as 20 hertz to as high as 20 000 hertz.

Humans are unable to hear sounds with a frequency above 20 000 hertz. High frequency sounds above 20 000 hertz are called ultrasounds.

The table 4.4 lists the upper and lower frequency limits for a number of animals.

| Animal | Frequency of lower limit of hearing (hertz) | Frequency of upper limit of hearing (hertz) |
|---|---|---|
| Cat | 30 | 45 000 |
| Dog | 20 | 30 000 |
| Human | 20 | 20 000 |
| Whale | 40 | 80 000 |

**Table 4.4**

## Ultrasound in medicine and in industry

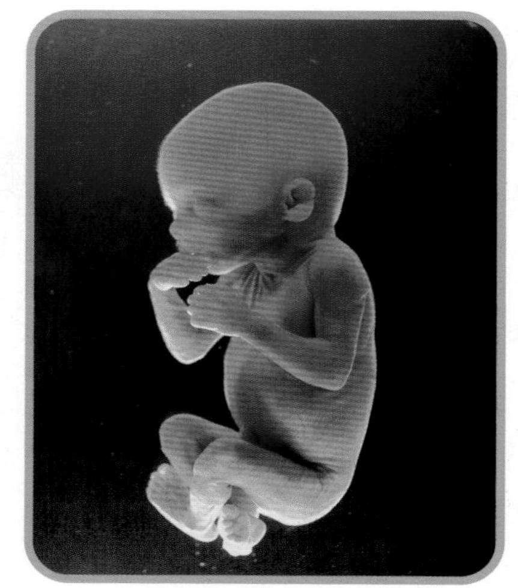
**Figure 4.17** *An unborn baby.*

Ultrasound can be used to take 'pictures' of the inside of the body. Figure 4.17 shows the image of a baby inside a mother's womb.

Ultrasound can also be used continuously to monitor movements within the body.

A three-dimensional picture can be obtained as the reflected signals take different times to return from objects at different distances from the ultrasound source.

Dentists use ultrasound — ask your dentist the next time you visit how ultrasound is used.

Ultrasound is also used in industry to 'see' inside solid materials. Metal parts on an aircraft flex and bend during use. These metal components can have tiny cracks inside them. During flexing and bending these cracks can get larger and this can cause the component to break — this is known as metal fatigue. The metal components can be tested using ultrasound so that a 'picture' of the inside of the solid metal component can be obtained (figure 4.18).

**Figure 4.18** *Metal being tested for cracks using ultrasound.*

## Measuring sound level

The human ear is a sensitive detector of sound and can be damaged by very loud sounds.

The loudness of sound is measured in **decibels** using a sound level meter.

In school you may be able to use a sound level meter to measure the loudness of some sounds.

Switch on the sound level meter and record the sound level readings for different sounds (point the detector at the source of the sound), such as a quiet area, a radio, a person talking quietly (1 metre from the sound level meter), a person talking loudly (1 metre from the sound level meter). Draw a table of your results.

## Noise pollution

Make a list of four noises that are unpleasant to you.

Noise is unwanted sound. It may be sound from traffic, from a neighbour's TV or radio, from machinery at work, and so on. Because it is unwanted, noise is a kind of pollution.

| Source of Sound | Sound level (decibels) |
|---|---|
| Disco, 1 metre from loudspeaker | 120 |
| Pneumatic drill at 5 metres | 100 |
| Alarm clock 0.5 metres from bedside | 80 |
| Normal conversation at 1 metre | 60 |
| Residential area at night | 40 |
| Quiet country lane | 20 |
| Silence (hearing threshold) | 0 |

**Table 4.5**

Excessively loud noises are unpleasant and some can cause damage to the hearing.

In factories, noisy work places near pneumatic drills and aircraft, in heavy vehicles or tractors the noise level can be over 100 dB. This can cause permanent damage to the ears with a serious loss of hearing ability. A 'ringing' sound heard after exposure to this level of sound is a warning sign. Ear protection — earplugs, earmuffs or a helmet — can be used to reduce the level of noise heard. They do this by preventing most of the sound vibrations reaching the ear.

During a normal working day, you should not be exposed to a noise level above 90 dB unless you have some type of protection for your ears.

## Key Points

- Sound can pass through solids, liquids and gases.
- Sound cannot pass through a vacuum.
- The normal range of humans is from 20 hertz to 20 000 hertz.
- High frequency sounds beyond the range of human hearing, above 20 000 hertz, are called ultrasounds.
- Ultrasound can be used to produce images of an unborn baby.
- Ultrasound can be used to produce images of cracks inside a metal.
- The loudness of a sound is measured in decibels.
- Sounds above 90 decibels can cause damage to hearing.

## Section Questions

**1** Copy and complete the following sentences using appropriate words from the list:
gases particles vacuum.

Sound waves from a vibrating object can travel through solids, liquids and ______. Solids, liquids and gases are made up of ______ which are made to vibrate by the sound waves. Sound cannot pass through a ______ since there are no particles to vibrate.

**2** Copy and complete the following sentences using appropriate words from the list:
damage frequencies decibel ultrasounds.

Most humans can hear sounds with ______ from about 20 hertz to 20 000 hertz. Sounds with a frequency above 20 000 hertz are called ______. The loudness of sound is measured on the ______ scale. Sounds above 90 decibels can cause long-term ______ to hearing.

**3** A dog owner cannot hear the sound from a whistle, although the dog does. Suggest a suitable value for frequency of the sound from the whistle.

**4** Table 4.6 shows the upper limit of hearing for a number of animals.
a) Present the information in the table as a bar chart.
b) Which of these animals would be able to hear a sound of frequency 35 000 hertz?

| Animal | Frequency of upper limit of hearing (hertz) |
|---|---|
| Cat | 45 000 |
| Dog | 30 000 |
| Human | 20 000 |
| Whale | 80 000 |

**Table 4.6**

**5** Ultrasound is important in both medical and industrial applications.
a) Give one medical use of ultrasound.
b) Give one non-medical use of ultrasound.

**6** Noise pollution can be a big problem to those people affected. Give two examples of noise pollution.

**7** What noise level, in decibels, can cause serious damage to hearing over a long period of time?

**8** What name is given to frequencies of sound beyond the range of human hearing?

**9** Engineers working near aeroplane jet engines wear ear protectors. Explain why the ear protectors are needed.

# Amplified sound

## After studying this section you should be able to...

1. State that the output signal from an amplifier has the same frequency but a bigger amplitude than the input signal.
2. State the function of the three major components needed to amplify speech (microphone, amplifier, loudspeaker).
3. Define the voltage gain of an amplifier in terms of input and output voltages.
4. Calculate voltage gain using: voltage gain $= \frac{\text{output voltage}}{\text{input voltage}}$.
5. Explain why your recorded voice sounds different to you.
6. State the advantages of a compact disc compared to a tape cassette.

## Amplified sound

An amplifier is an electrical device which makes electrical signals bigger.

The effect of an amplifier on an electrical signal is shown in figure 4.19.

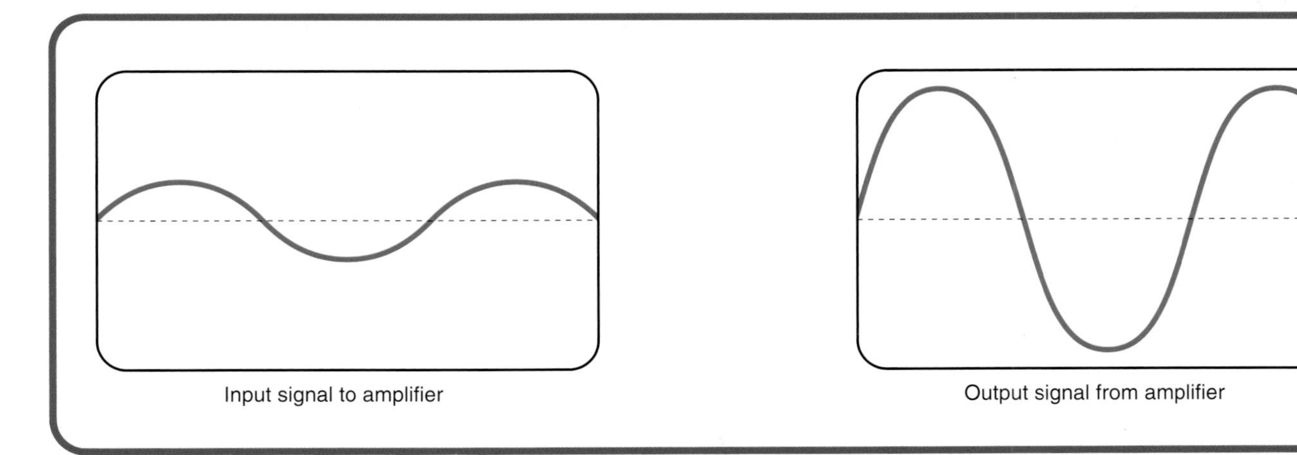


**Figure 4.19**

Notice that the input signal to the amplifier is smaller than the output signal.

The output signal from an amplifier can be made very much bigger than the input signal – the energy required to do this comes from the battery or mains supply to the amplifier.

The output frequency from the amplifier is the same as the input frequency.

In appliances producing sound the amplifier is generally the volume control.

Some examples of appliances which contain an amplifier are a television, a hi-fi and a radio.

Some students in the school are holding a rock concert during the lunch break. You are given the task of organising a sound system for a large room. What are the major components required to do this?

The voices/music must be made louder.

To do this the sound must be changed into an electrical signal by a microphone. This electrical signal can be made bigger using an amplifier. The bigger or amplified electrical signal is then changed back into sound by a loudspeaker. A block diagram of the components is shown in figure 4.20.

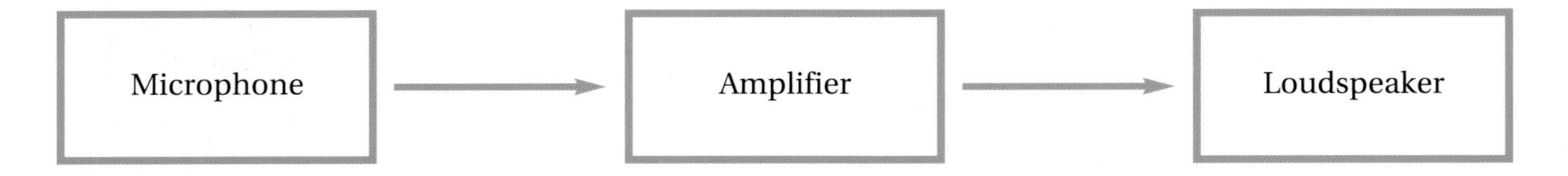


| Name of Component | What it does |
|---|---|
| Microphone | Changes the sound into an electrical signal |
| Amplifier | Makes the electrical signal bigger |
| Loudspeaker | Changes the amplified electrical signal into sound |

**Figure 4.20**

The input signal to and the output signal from an amplifier are shown in figure 4.21.

Notice that the frequency of the output signal is the same as that of the input signal.

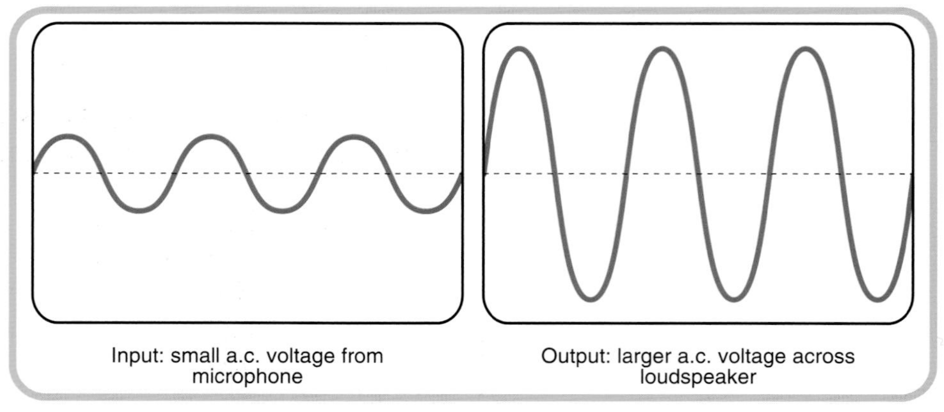


**Figure 4.21**

## Voltage gain

The input and output voltages from an amplifier are shown on the voltmeters in figure 4.22.

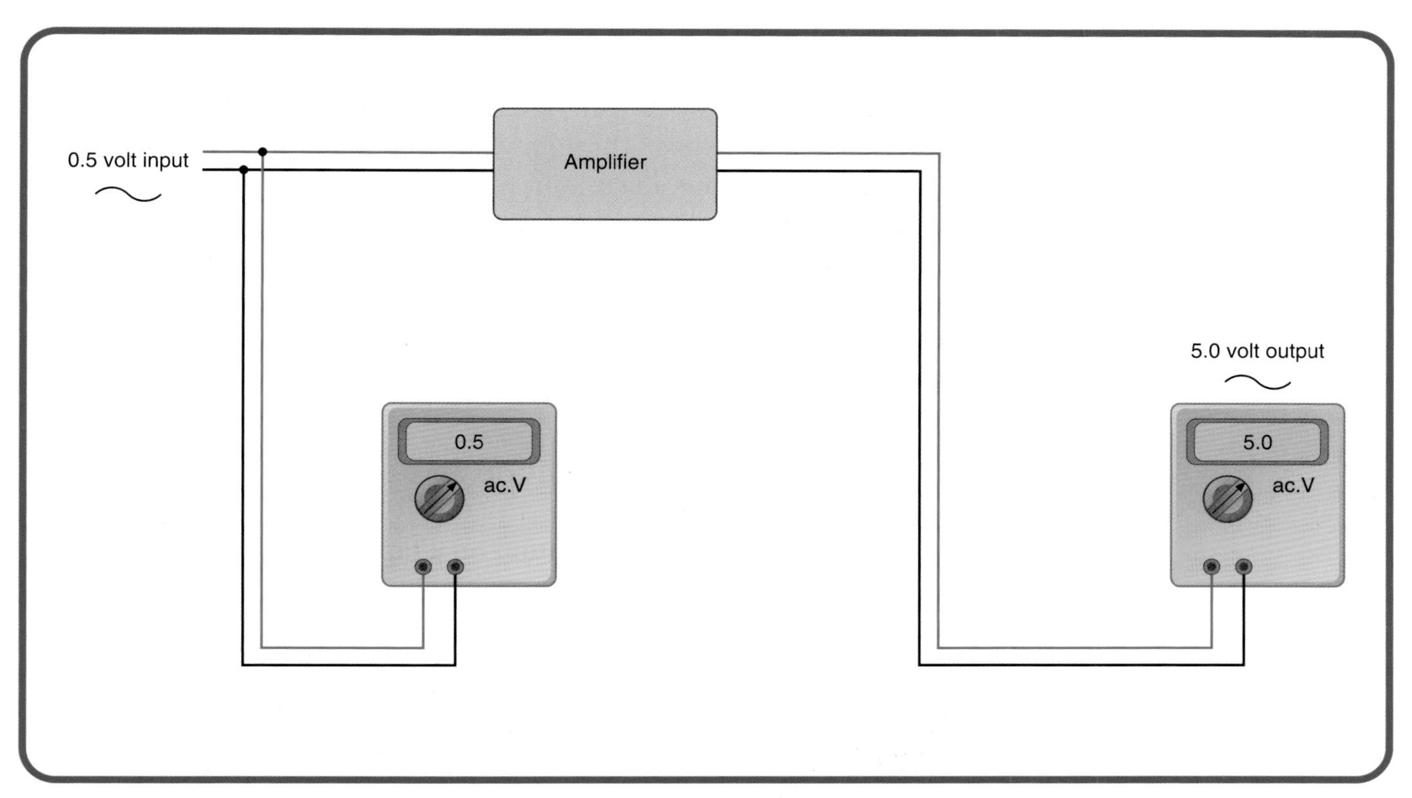


**Figure 4.22**

The input voltage to the amplifier is 0.5 V. The output voltage from the amplifier is 5.0 V, which is 10 times bigger than the input voltage. The amplifier has made the output voltage 10 times bigger than the input voltage. The voltage gain of the amplifier is 10.

$$\text{Voltage gain} = \frac{\text{output voltage}}{\text{input voltage}}.$$

### Examples

The input voltage to an amplifier is 0.03 V. The output voltage from the amplifier is 6.0 V. What is the voltage gain of the amplifier?

$$\begin{aligned}\text{Voltage gain} &= \frac{\text{output voltage}}{\text{input voltage}}\\ &= \frac{6.0}{0.03}\\ &= 200.\end{aligned}$$

The output signal from an audio amplifier has the same frequency as the input signal.

At home or at school record your voice onto a tape recorder. Play the tape back. How does your voice sound to you?

When you speak you hear the sounds that you make. The sound waves from your voice pass through the air, but also travel through the bones and tissue of your head to your ears. The sound waves travel at a higher speed through your head than through the air. When you hear your voice, your ears are picking up the sound that has travelled through your head as well as the same sound that has travelled through the air. This effect produces a slightly different sound compared to what other people hear. However, when the tape of your voice is played, you only hear the sound that travels through the air – this sounds very strange to you.

## Compact disc versus tape cassette

A tape cassette stores information on a tape using magnetism. A compact disc stores information digitally — using a series of logic 0s and logic 1s.

Compared to tape cassettes, compact discs:
- can store more information,
- give a higher quality recording,
- are less likely to be damaged — the tape in a cassette can get snagged in the recorder and the tape can be damaged.

### Key Points

- An amplifier makes electrical signals larger.
- The output signal from an amplifier has the same frequency but a bigger amplitude than the input signal.
- Voltage gain = $\dfrac{\text{output voltage}}{\text{input voltage}}$.
- Compact discs can store more information and can give a higher quality recording than cassette tapes.

## Section Questions

**1** Copy and complete the following sentences using appropriate words from the list:

process    input    same

An amplifier is a ______ device which makes the output signal bigger than the ______ signal. The frequency of the output signal remains the ______ as the input signal.

**2** A number of input, process and output components are listed in table 4.7.

| Input | Process | Output |
|---|---|---|
| Solar cell | AND gate | Electric motor |
| Switch | Amplifier | Lamp |
| Microphone | NOT gate | Loudspeaker |

**Table 4.7**

Which of the above components are required to be able to amplify a sound?

**3** A number of electrical devices are listed below.

electric kettle    radio    television
vacuum cleaner

From the above list, write down the names of two devices which contain an amplifier.

**4** Copy and complete table 4.8 for the voltage gain of each amplifier.

| Name of amplifier | Input voltage (volts) | Output voltage (volts) | Voltage gain |
|---|---|---|---|
| AEB | 0.01 | 3.0 | |
| DMC | 0.05 | 2.5 | |
| RWG | 4.0 | 6.0 | |

**Table 4.8**

**5** The input voltage to an amplifier is 0.12 volts. The output voltage from the amplifier is 6.0 volts. Calculate the voltage gain of this amplifier.

**6** The input voltage to and output voltage from an amplifier are displayed on the voltmeters shown in figure 4.23.

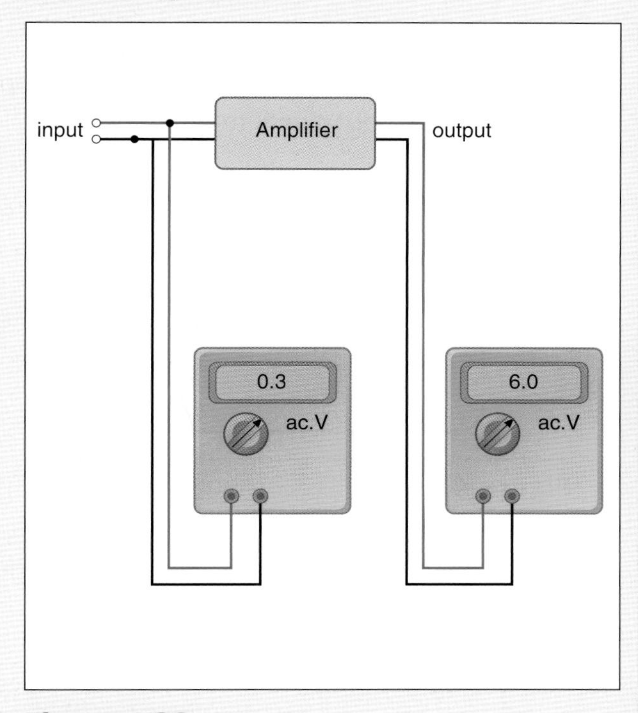


**Figure 4.23**

Calculate the voltage gain of the amplifier.

**7** Part of the data sheet for an amplifier is shown below.

Input voltage = 0.01 volts
Minimum output voltage = 1.0 volts
Maximum output voltage = 4.0 volts
Maximum voltage gain = 500

Has the information on the data sheet for the maximum voltage gain been printed correctly? You must show your working clearly.

**8** You record your voice on a tape recorder and then replay it back. Why does your recorded voice sound different to you?

**9** State two advantages of a compact disc compared to a tape cassette.

**10** Part of a baby alarm consists of three components: a microphone, an amplifier and a loudspeaker.

State the function of each of these components in the baby alarm.

**11** Figure 4.24 shows the trace on an oscilloscope screen of the input signal to an amplifier.

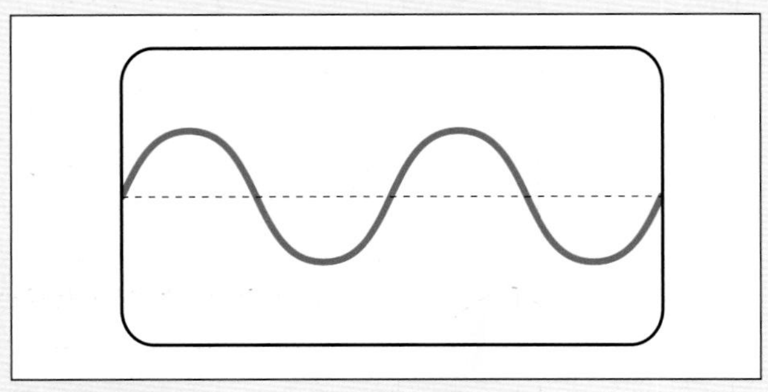

**Figure 4.24**

The controls of the oscilloscope are not changed. The oscilloscope is connected to the output of an amplifier.

Which diagram in figure 4.25 shows the trace seen on the screen of the oscilloscope?

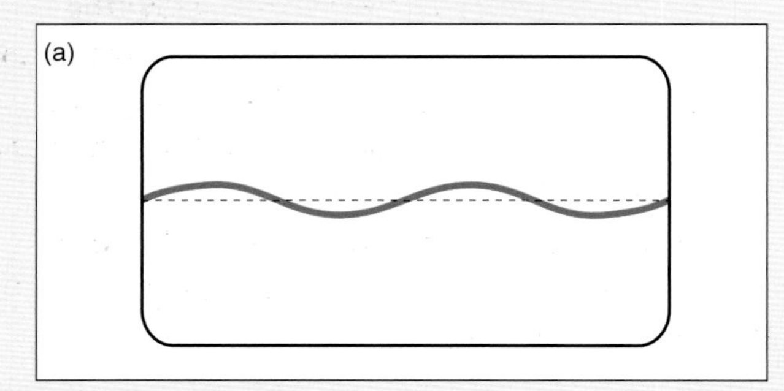


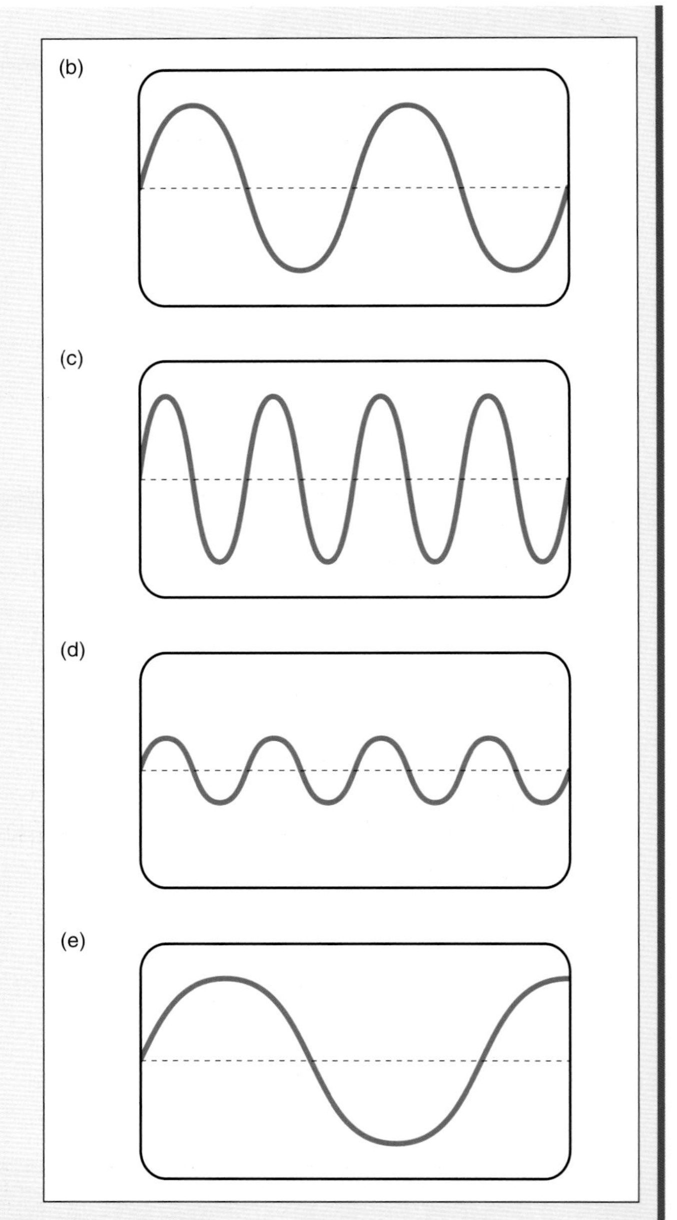


**Figure 4.25**

## Examination Style Questions

1

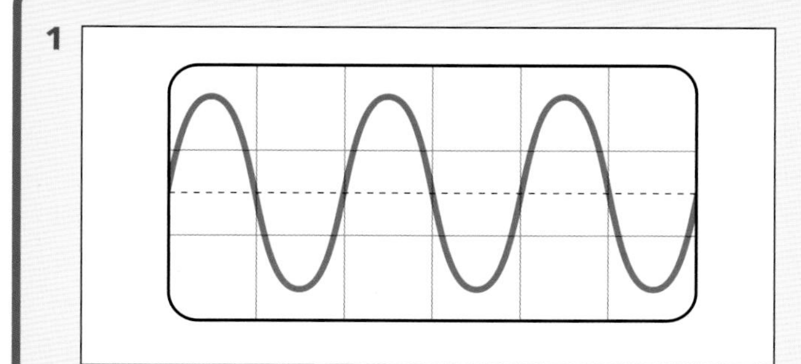

Figure 4.26

a) A student plays a musical note from a violin into a microphone. The microphone is connected to an oscilloscope. The student examines the pattern displayed on the oscilloscope screen. The pattern is shown in figure 4.26.

(i) The student plays a second note into the microphone. The controls on the oscilloscope are unchanged. This note has the same loudness but a lower frequency than the first note.

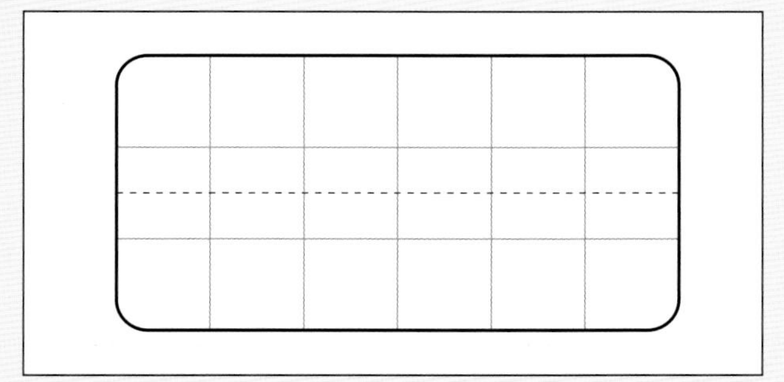

Figure 4.27

Copy the diagram shown in figure 4.27. Complete your diagram by drawing the pattern displayed on the oscilloscope screen produced by the second note.

(ii) State one alteration the student could make to the string to produce a note with a lower frequency.

b) The student plays a note with a frequency of 512 hertz. Calculate the frequency of a note one octave lower.

2

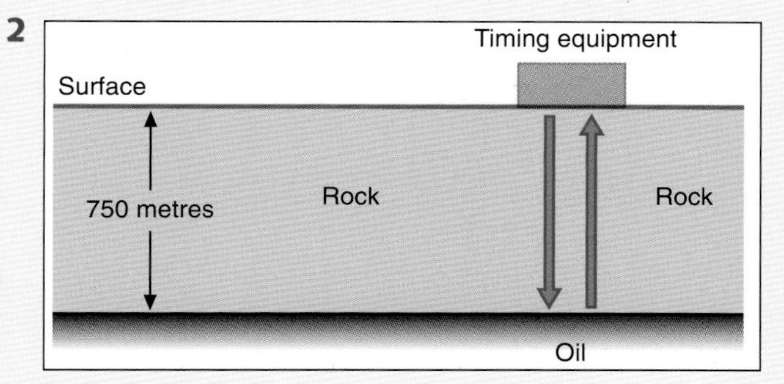


Figure 4.28

A physicist uses sound waves in the search for oil. An explosion is produced on the surface. The sound from the explosion travels through the rock and is reflected from the top of the oil as shown in figure 4.28.

The top of the oil is 750 metres below the timing equipment. The time between the sound being produced and received is 0.3 seconds.

a) How far does the sound wave travel through the rock in 0.3 seconds?

b) Calculate the speed of sound in the rock.

c) When the sound is produced the physicist has to wear ear protectors.

Explain why the ear protectors are needed.

**3** Copy and complete the crossword on sound.

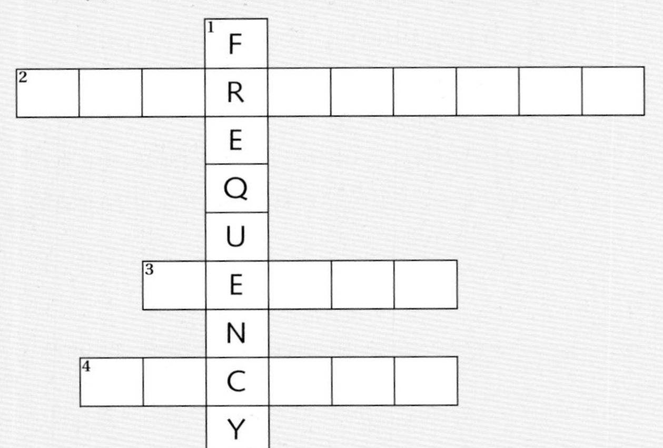


Across 2. The name given to frequencies of sound beyond the range of human hearing.

3. The unit of frequency.

4. Sound cannot pass through this.

Down 1. The number of waves in 1 second.

**4** The output signal from an amplifier is displayed on the screen of an oscilloscope as shown in figure 4.29.

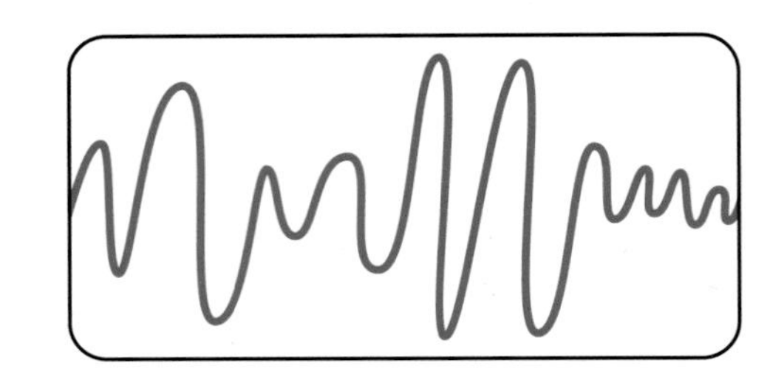

**Figure 4.29**

a) Is this an analogue or a digital signal?

b) The voltage of the input signal to the amplifier is 0.05 volts. The voltage of the output signal from the amplifier is 3.0 volts. Calculate the voltage gain of the amplifier.

c) Another amplifier has a voltage gain of 10. The frequency of the input signal to this amplifier is 500 hertz.

What is the frequency of the output signal from the amplifier?

# Movement

You opened this book using a movement, but what is a force or a speed or even acceleration? What happens when forces act? Why are some collisions more dangerous than others? This section looks at forces in action, from cars to aircraft to sports.

# Forces

## After studying this section you should be able to...

1. Describe how to use a Newton balance to measure force.
2. State that weight is a force and is the Earth's pull on the object.
3. Calculate weight using weight = 10 x mass.
4. State that the force of friction can oppose the motion of an object.
5. Describe one way in which the force of friction can be increased.
6. Describe one way in which the force of friction can be decreased.
7. State the benefits of streamlining an object
8. Describe two features of a car which improve streamlining.
9. State that equal forces acting in opposite directions on an object are called balanced forces.
10. Identify situations where the forces acting on an object are (a) balanced, (b) not balanced.
11. State that when the forces acting on an object are balanced the movement of the object does not change.
12. State that when the forces acting on an object are not balanced the speed and/or direction of the object changes.

## Forces—*everywhere*

Forces exist in lots of places. There is one acting on all of us due to gravity which keeps us on the ground! There are forces between magnets, which attract or repel (figure 5.1). Forces, like pushes and pulls, can have different effects on objects.

**Figure 5.1**

**Figure 5.2**

Figure 5.2 shows a force acting on a ball.

- The force can change the **speed** of the ball.
- The force can change the **direction** of the ball.
- The force can change the **shape** of the ball, by squashing it.

These effects depend on the size of the force on the object.

## Measuring force

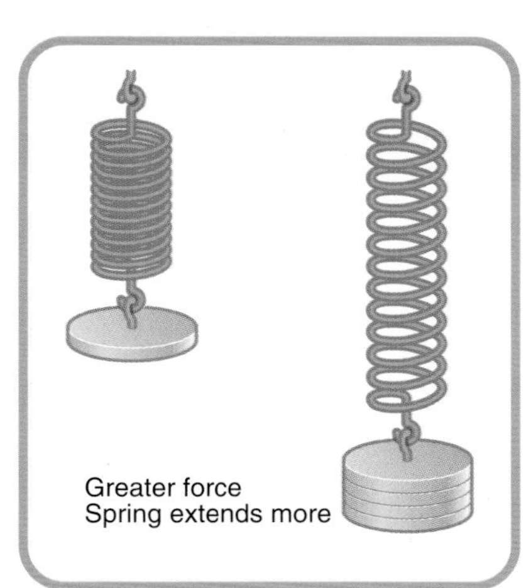


**Figure 5.3**

Springs can be used to measure force.

1. A spring stretches evenly — a bigger force applied to the spring causes the spring to stretch even further (figure 5.3).
2. A spring returns to its original length when the force is removed.

We measure forces using a Newton balance (figure 5.4).

**Figure 5.4** *A newton balance.*

This uses a spring inside a tube and a scale on the outside which is in newtons or N.

The Newton balance is named after the famous physicist Sir Isaac Newton. If you add different masses to the balance then you can weigh different objects. If you pull a trolley then you can measure the pulling force. Forces can go from a few newtons to pull a trolley in the laboratory to hundreds of thousands of newtons to move an aircraft (figure 5.5).

**Figure 5.5**

## Weight

Weight is the force due to gravity. The force of gravity happens between any objects (us!) and the Earth. It is the pull of the earth on us.

If there was no force of gravity then we would become weightless. This has never happened because we would have to be deep in space away from any planets.

If we could travel to other planets then we would weigh different amounts since the pull of other planets on us is different from that of the Earth.

## Calculating weight

To calculate weight you need to know your mass. This is the number and type of particles that you are made from and this does not depend on gravity like weight does. It is measured in kilograms. Most people confuse mass and weight. We normally talk about weight as 50 kilograms but this is wrong. Only mass can be measured in kilograms because your mass cannot change if you travel to another planet with a different force of gravity. Remember that:

**Mass is in kilograms and weight is in newtons.**

If you want to calculate your weight then you do it as:

Weight = mass x 10.

### Examples

Sheila has a mass of 48 kilograms. What does she weigh on Earth?

Solution
Weight = mass x 10
= 48 x 10
= 480 newtons.

### Examples

Fred has a mass of 67 kilograms. What is his weight on Earth?

Solution
Weight = mass x 10
= 67 x 10
= 670 newtons.

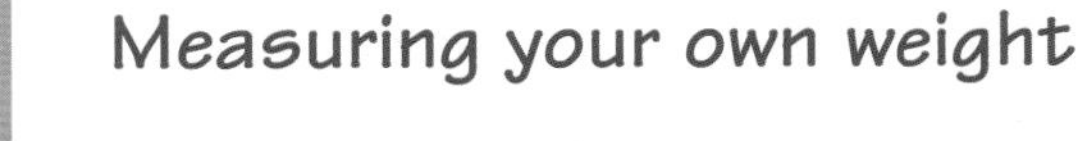

## Measuring your own weight

You can measure your own weight by using a set of scales which are in newtons rather than kilograms. Stand on the scales and note what your weight is. Check that if you divide by 10 then you get your mass in kilograms (figure 5.6).

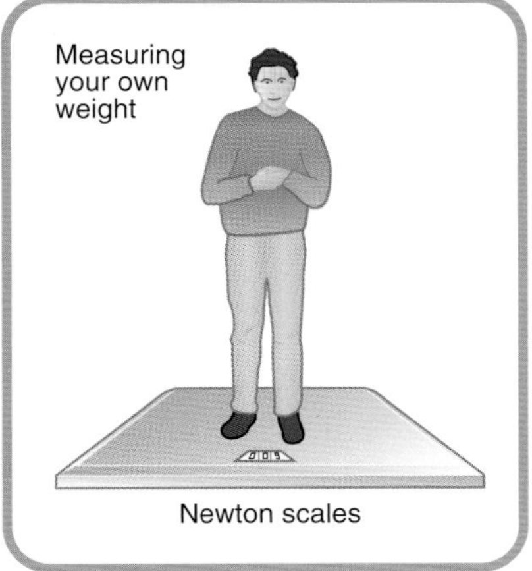


**Figure 5.6**

## Frictional forces

Cars can slow down due to forces acting on them. These forces can be due to the road surface and the tyres or due to the brakes or even the air passing over the moving car. The force that tries to oppose the motion is called the force of friction. A frictional force always acts when two surfaces are rubbing against each other. Friction can oppose any motion. The force of friction will depend on the type of surface. Friction will try to slow down an object like a car no matter which direction it is travelling in along a road (figure 5.7).

**Figure 5.7**

## Increasing friction

If we consider a car then there are several things that make the force of friction acting on the car bigger:

- making the front area of the car larger,
- using the brakes on the car to slow it down,
- fitting a roof rack to a car( figure 5.8).

**Figure 5.8**

## Car brakes

The task of the brakes is to slow down the car. When the brake pedal is pressed it causes the brake shoes to be pushed against the drum. Friction then stops the wheel going round (figure 5.9).

**Figure 5.9** *The brake shoes of a car.*

## Reducing friction

The frictional force between two surfaces moving against each other can be reduced by lubricating the surfaces. This generally means that oil can be placed in between two metal surfaces. This happens in car engines and reduces wear on the engine since the metal parts are not actually meeting each other, but have a thin layer of oil between them. If there is oil on the road then the car might skid since there is little friction between the road surface and the tyre (figure 5.10).

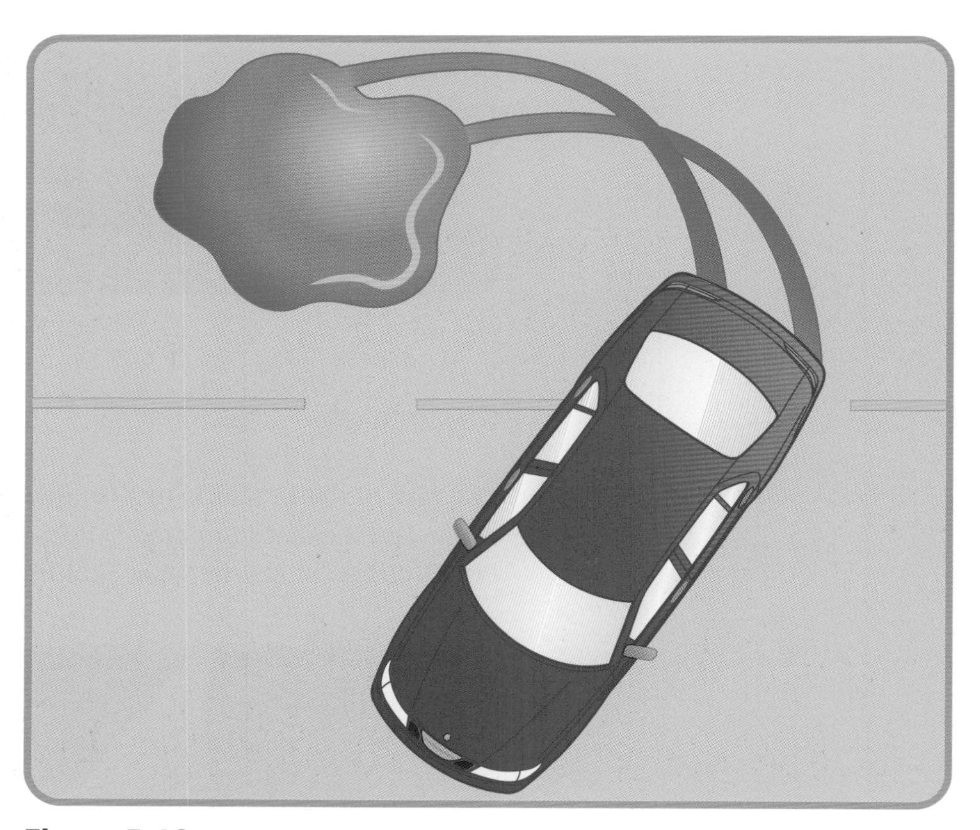

**Figure 5.10**

Skis can be waxed so that they have little friction when they come into contact with the snow (figure 5.11).

**Figure 5.11**

## Streamlining

Making a gas or liquid flow easily over a surface is called streamlining.

Modern cars are designed to offer as little resistance (drag) to the air as possible. This is the friction of air on the car. To reduce this air friction the designers try to streamline the vehicle in a variety of ways. To help design the vehicle it is placed in wind tunnels and smoke is allowed to flow over it, as shown in figure 5.12. The smoother the flow of air over the car the better the streamlining.

**Figure 5.12** *Airflow over a car.*

**Figure 5.13** *Low profile tyres.*

The air friction on a car can be reduced in a number of ways:

- reducing the front area of the car,
- not carrying a roof rack,
- having door mirrors instead of wing mirrors,
- having a smooth round body shape,
- having aerials made as part of the car windows.

In racing cars air friction can be reduced in a number of ways:

- a wing at the back to help produce a downward force,
- thinner tyres, called low profile tyres, also help (figure 5.13).

## Humans

When you try to move fast through water as a swimmer then it helps to have as smooth an outline as possible. The modern design of swimwear produces a smooth shape so that the disturbance of the water is reduced (figure 5.14). In the same way cyclists wear close fitting clothes to allow the air to flow over them easily and to make them move faster (figure 5.15).

Figure 5.14

Figure 5.15

## Balanced and unbalanced forces

When two or more forces act on an object the total effect depends on their size and direction.

### Balanced forces

Balanced forces act on an object:

- they are equal in size,
- they act in opposite directions.
- they cancel each other out and so no overall force acts on the object (figure 5.16).

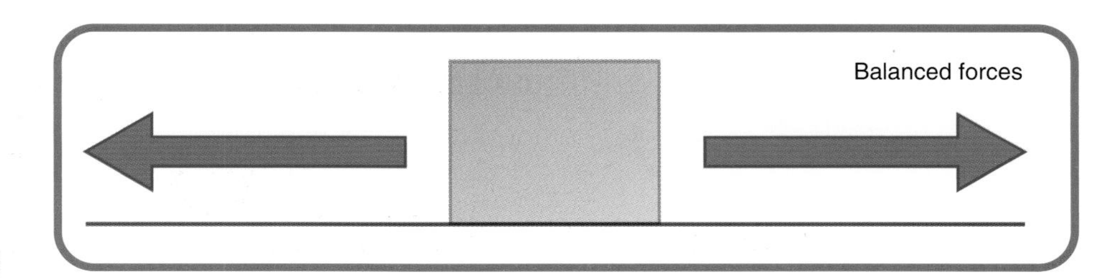


Figure 5.16

**Figure 5.17**

When forces A and B in figure 5.17 are the same size, the same force is applied to each side of the vehicle. The forces are balanced. If the vehicle is not moving it will stay where it is. If the vehicle is moving, it will continue to move at a constant speed in a straight line.

When the engine force A is greater than the air resistance force B, the car will accelerate to the right.

Examples of balanced forces:

- A ship moving at a steady speed through the water (figure 5.18). The resistance force of the water balances out the engine force.

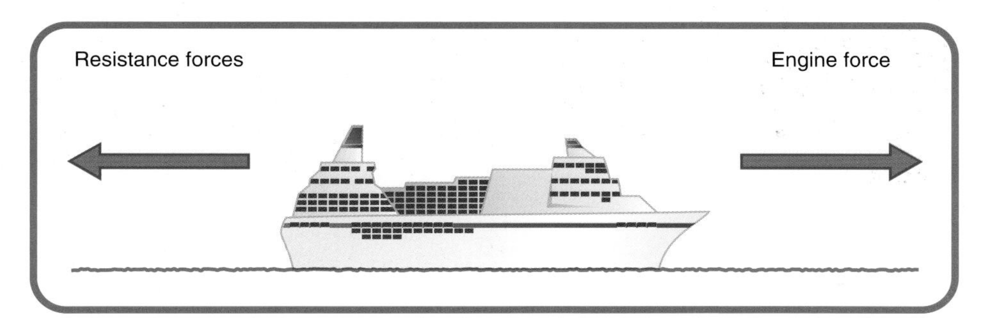


**Figure 5.18**

- A car is moving at a constant speed along a level road. No matter how hard the driver tries to accelerate the car does not increase its speed (figure 5.19). The air resistance force has balanced the force produced by the engine.

**Figure 5.19**

- When you wish to dislodge sauce from the bottom of a bottle you turn it upside down and throw it downwards very fast and then bring it to halt. The bottle has stopped, but the sauce keeps moving at a constant speed in a straight line. The forces on the sauce are balanced.
- If you become a sky diver then you will dive out of an aircraft and fall to the ground. In the first stage you will not open your parachute. As you do this the forces acting on you will eventually balance. Your weight acting down is being balanced by the air resistance forces acting upwards. By this time your speed is about 60 metres per second or about 140 miles per hour! (figure 5.20). You will now open your parachute and your speed will change to about 10 metres per second and you can land safely (figure 5.21).

**Figure 5.20**

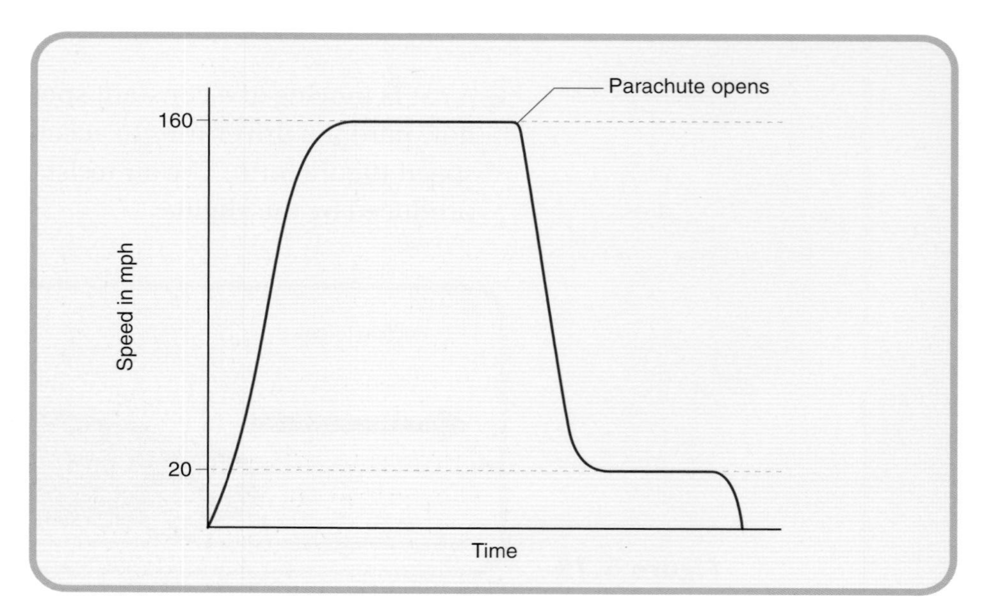


**Figure 5.21**

### Unbalanced forces

When one force is larger than the other then the object can either:

- change direction,
- change speed.

One or both of these effects can happen at once.

An aircraft produces more engine force (thrust) than the drag force. The aircraft will increase its speed and the direction it moves in could change (figure 5.22).

In a car, if the braking force is larger than the engine force then the car will slow down.

**Figure 5.22**

## Key Points

- A Newton balance is used to measure force in newtons.
- Weight is a force and is the Earth's pull on the object.
- Weight = 10 x mass.
- The force of friction can oppose the motion of an object.
- The force of friction can be increased by a rough surface.
- The force of friction can be decreased by using oil between two surfaces.
- Streamlining reduces the effect of air friction on an object.
- Two features of a car which improve streamlining are to have a very smooth shape and have no objects which stick out, such as wing mirrors.
- Forces of the same size acting in opposite directions on an object are called balanced forces.
- When the forces acting on an object are balanced the movement of the object does not change.

## Section Questions

**1** A wooden crate is dragged across a room. The force to do this can be measured.

a) What is used to measure the force?

b) What is the unit of force?

**2** When you are on the Earth, there is a force acting on you called weight. What causes this force on you?

**3** Two pupils have different masses. Margaret has a mass of 50 kilograms and Tom has a mass of 68 kilograms. Calculate the weight of Margaret and Tom on Earth.

**4** When a box is pulled across a floor, there is a force between the box and the floor. This force tries to stop them moving. What is this force called?

**5** State one way in which the force of friction between two surfaces can be (a) increased, (b) decreased.

**6** A new aircraft is designed to reduce the air friction acting on it at high speed. What is the name for the design change which reduces the effect of air friction?

**7** A car is being designed to travel at high speeds. State two ways in which the design of the car would allow the air to flow over it more easily.

**8** A tug of war has one team pulling with a force of 800 newtons. The rope does not move.

a) What is the size of the force being exerted by the other team?

b) What name is given to this pair of forces?

**9**

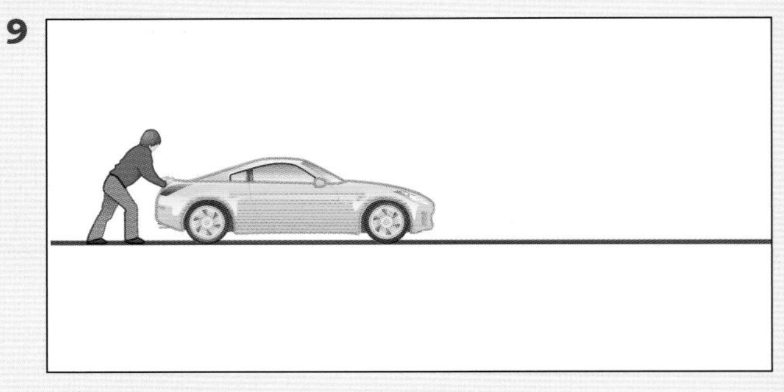

**Figure 5.23**

A car is being pushed along the road to start it on a cold day (figure 5.23). The force exerted by the two people is 200 newtons. The force of friction on the car is 200 newtons.

a) Will the car speed up, slow down or stay at a constant speed?

b) Explain your answer.

**10** Why is braking a car on a wet road more difficult than on a dry road?

# Speed and acceleration

## After studying this section you should be able to...

1 Describe how to measure average speed.

2 Calculate average speed using average speed $= \frac{\text{distance}}{\text{time}}$.

3 Describe how to measure instantaneous speed.

4 Describe the term acceleration.

## Speed

Speed is the distance travelled by an object in 1 second:

$$\text{speed} = \frac{\text{distance}}{\text{time}}.$$

A boy travels a distance of 200 metres from A to B in a time of 100 seconds. The average speed of the boy is 2 metres per second. However the boy's speed may change throughout the journey. He might be slower than 2 metres per second at one stage and faster than 2 metres per second at another stage (figure 5.24).

$$\text{Average speed} = \frac{\text{distance}}{\text{total time taken}}.$$

Speed and average speed have the same unit of metres per second.

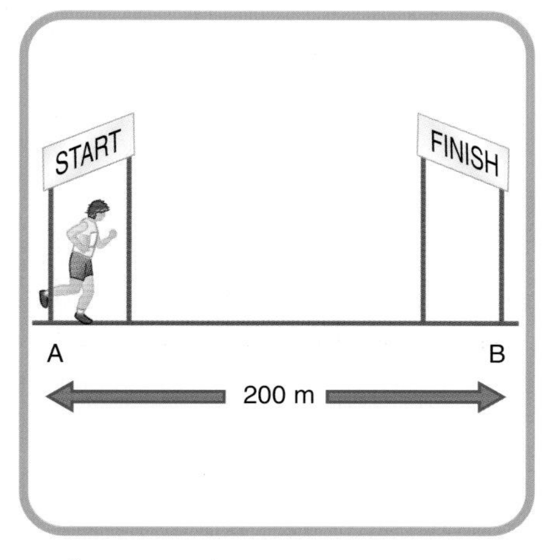


Figure 5.24

## Measuring average speed

Average speed can be measured using a stopwatch, trolley and measuring tape.

The distance is measured between two points A and B a few metres apart on the ground. This can be done using a measuring tape. The time for the journey is measured by starting a stopwatch when the trolley reaches A and stopping the watch when it reaches B (figure 5.25).

Figure 5.25

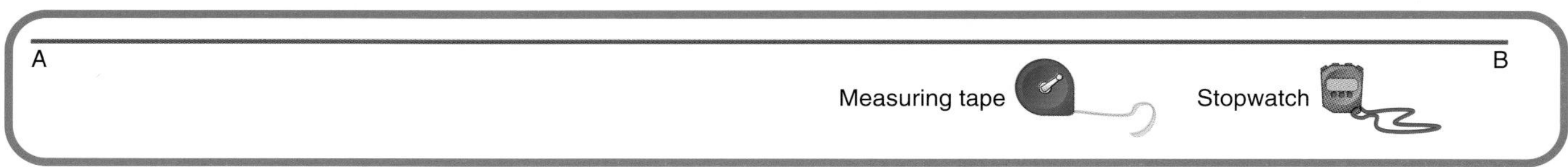

### Examples

During a sports match a player runs a distance of 15 metres in 2.5 seconds. Calculate his average speed.

Solution

$$\text{Average speed} = \frac{\text{distance}}{\text{time}} = \frac{d}{t}$$

$$= \frac{15}{2.5}$$

$$= 6 \text{ metres per second.}$$

## Instantaneous speed and average speed

If we could measure very small amounts of time, for example hundredths or even thousandths of a second, then we could measure the speed of any object just at that moment in time.

Instantaneous speed is the speed of an object at a particular time (instant).

In a car journey the speed of the car will change several times due to the traffic conditions. At the start the car will speed up then slow down as the traffic around it slows down. It may come to a halt at a set of traffic lights. After that the car speeds up again. If the speed is measured at one point during the journey then that would be the instantaneous speed (figure 5.26).

**Figure 5.26**

## Measuring instantaneous speed using a computer and light gate

The computer uses an internal clock which allows very small amounts of time to be measured. This allows the calculation of instantaneous speed if a distance is measured.

The computer starts timing when a light beam from the light gate is cut by the card. The computer stops when the card has passed the light beam. The time taken for the card to pass through the beam is recorded by the computer (figure 5.27).

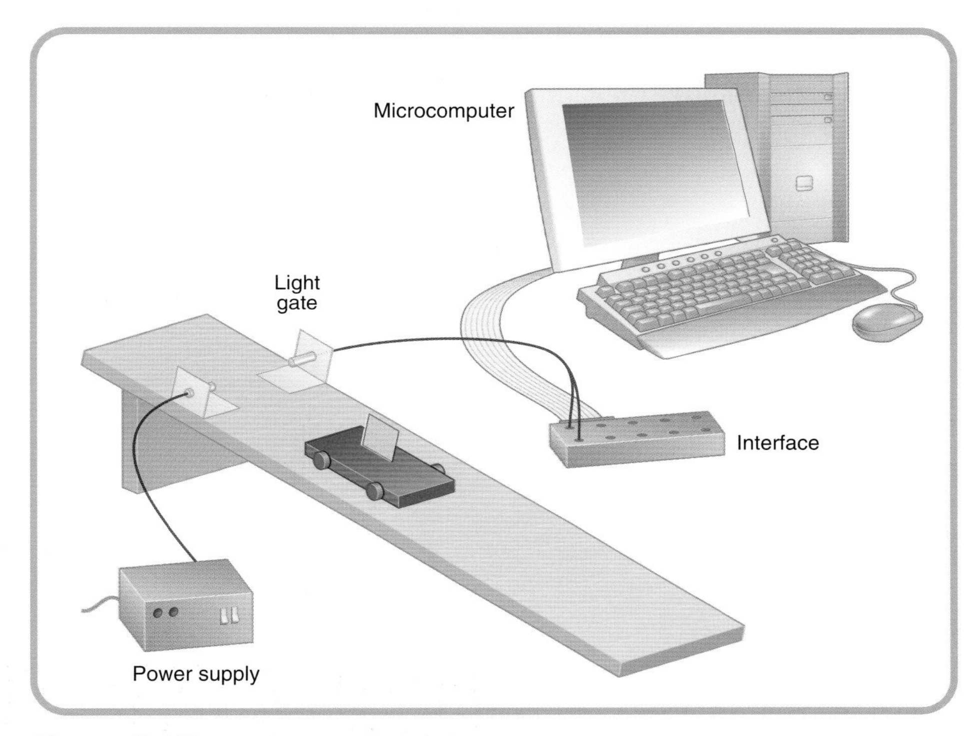


**Figure 5.27**

The computer, already pre-programmed with the length of the card, calculates the speed of the vehicle using:

$$\text{speed} = \frac{\text{length of card}}{\text{time to break beam of light}}.$$

Typical results might be:

Length of card = 10 cm = 0.1 metres

Time measured by computer = 0.05 seconds

$$\text{Speed} = \frac{0.1}{0.05}$$
$$= 2.0 \text{ metres per second}$$

## Physics for you

### Measuring car speeds

There are several ways that the speed of moving cars can be measured.

To catch drivers who were breaking the speed limit, the police used a similar method to the one in the lab. A distance was measured between two points and cars were timed as they went between the two points. There was a table of measurements which allowed the police to calculate the speed of the car.

With new technology there are now several ways to detect speeding drivers.

### VASCAR

This is the speed system used in police cars.

There are two switches, the first one measures the time taken to travel between two markers. The second switch measures the distance between the two markers as covered by the moving police car.

It is essentially a stopwatch method and calculates average speed. There is no need for the police car to travel at the same speed or even in the same direction as the moving car (figure 5.28).

**Figure 5.28** *VASCAR system.*

### Speed guns

These are pointed at a vehicle and work by sending out radio waves. When the radio waves are reflected back from the car there is a change in frequency. The change of frequency is used to provide information about the speed of the car (figure 5.29).

**Figure 5.29** *A speed gun.*

### Speed cameras

A sensor in the camera detects a speeding car. After the car passes the camera it takes two photos half a second apart. There are lines two metres apart on the road. The distance between a number of marks and the time taken are known. The speed can then be calculated (figure 5.30).

The cameras are often called Gatso traffic cameras since they were developed by Maurice Gatsonides. He was a former Dutch racing driver.

**Figure 5.30**

## Acceleration

When the speed of a car increases, it is said to be accelerating. Acceleration is the change in speed in 1 second.

Table 5.1 provides some information about two vehicles.

| Name of vehicle | Time in seconds to go from 0 to 60 kilometres per hour | Top speed in kilometres per hour |
|---|---|---|
| Fleet | 6 | 270 |
| Arrow | 8 | 300 |

**Table 5.1**

Table 5.1 shows that:

- Arrow can travel faster than Fleet.
- Fleet has a larger acceleration than Arrow. This is because the time taken for the speed of Fleet to change by 60 kilometres per hour is shorter. The less time it takes for the speed of the car to change, the bigger the acceleration of the car.
- If the speed of an object changes, the object is said to be accelerating.

### Key Points

- Average speed $= \frac{\text{distance}}{\text{time}}$
- Instantaneous speed is calculated the same way as average speed but the time interval is very short.
- In both cases the unit is metres per second.
- Acceleration means that the speed of the object is increasing.

## Section Questions

**1** Calculate the average speed of a car which travels 2400 metres in 120 seconds.

**2** A car travels 45 metres in 9 seconds. Calculate the average speed of the car.

**3** An aircraft travels 1000 metres in 125 seconds. Calculate its average speed.

**4** A bicycle is travelling along a cycle track. Describe how you would measure the instantaneous speed of the cycle.

**5** A car magazine states that 'This car shows great acceleration'. Explain what is meant by the term acceleration.

**6** Table 5.2 gives details of three cars.

| Car name | Maximum speed in miles per hour | Time in seconds to go from 0 to 60 miles per hour |
|---|---|---|
| Argyle | 96.4 | 12.0 |
| Beardmore | 94.9 | 13.7 |
| Caledonian | 96.3 | 12.3 |

**Table 5.2**

a) Which car has the greatest maximum speed?
b) Which car has the largest acceleration?

# Moving objects

## After studying this section you should be able to...

1. Describe how the effect of a collision increases with the mass and speed of the objects involved.
2. State that the change in speed of an object increases with (a) the size of the force acting on it, (b) the time the force acts.
3. State that the range of a ball thrown at an angle is affected by (a) the speed of the throw, (b) the angle of the throw.
4. State that the height that a ball rebounds on hitting a surface is affected by (a) the speed on impact, (b) the material of the surface, (c) the material of the ball.

### Collisions

The most dangerous situation for car driving is when two cars collide (figure 5.31).

**Figure 5.31**

In a road safety laboratory the following experiments are carried out:

1. A car X of mass 1000 kilograms is parked at the side of the road. Another car Y of mass 1000 kilograms is travelling at 10 metres per second when it crashes into the first car. The two cars lock together and move off at 5 metres per second (figure 5.32).
2. A car Q is moving at 15 metres per second when it collides with car P. After collision the two cars lock together and move at 7.5 metres per second (figure 5.33(a)).
3. Car R is sitting at the side of the road. Car S has a mass of 1500 kg and is travelling at 10 metres per second when it collides with car R. After the collision the cars lock together and move off at 6 metres per second (figure 5.33(b)).

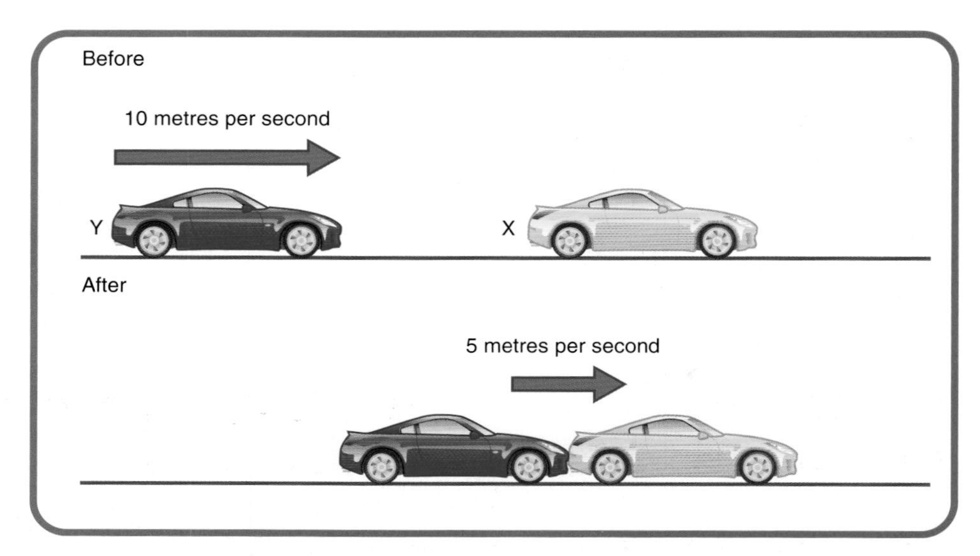


**Figure 5.32**

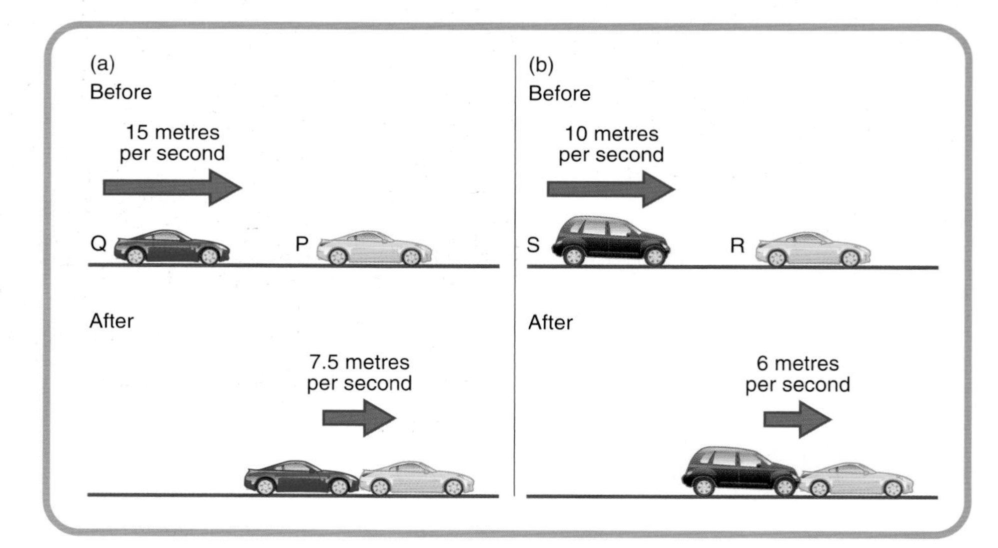


**Figure 5.33**

This means that in a collision between a moving car and a stationary car:

- The faster the moving car, the faster the two cars move off.
- The larger the mass of the car which is moving then the faster the two cars move off.
- Increasing the speed or the mass of the moving car before the collision means that both cars move off faster after the collision.

## Forces and impacts

During a tennis game a ball is struck with a racket and goes off at an angle.

The purpose of hitting the ball is to make it change its speed and possibly direction (figure 5.34).

**Figure 5.34**

The ball changes shape during the impact due to the forces acting on it (figure 5.35).

- The harder the ball is struck the faster it will move.
- The longer the ball is in contact with the racket the faster the ball will move.

**Figure 5.35**

In a baseball game the striker tries to hit the ball as hard as he can so that the ball moves off very fast (figure 5.36).

**Figure 5.36**

## Range of throw

Success in many sports uses the ideas of motion.

Some sports involve throwing away objects. Typical sports are basketball, baseball, cricket, football, shot put, hammer, discus, javelin, golf, volleyball and tennis (figure 5.37).

Figure 5.37

Figure 5.38

Sometimes the human body is thrown. Typical sports are long jump, gymnastics, figure skating, diving and ski jumping (figure 5.38).

The range of an object is how far it travels when it is thrown.

If you want to throw a ball or a javelin as far as you can:

- Increase the speed of throw. The faster an object is thrown the greater the range (figure 5.39).

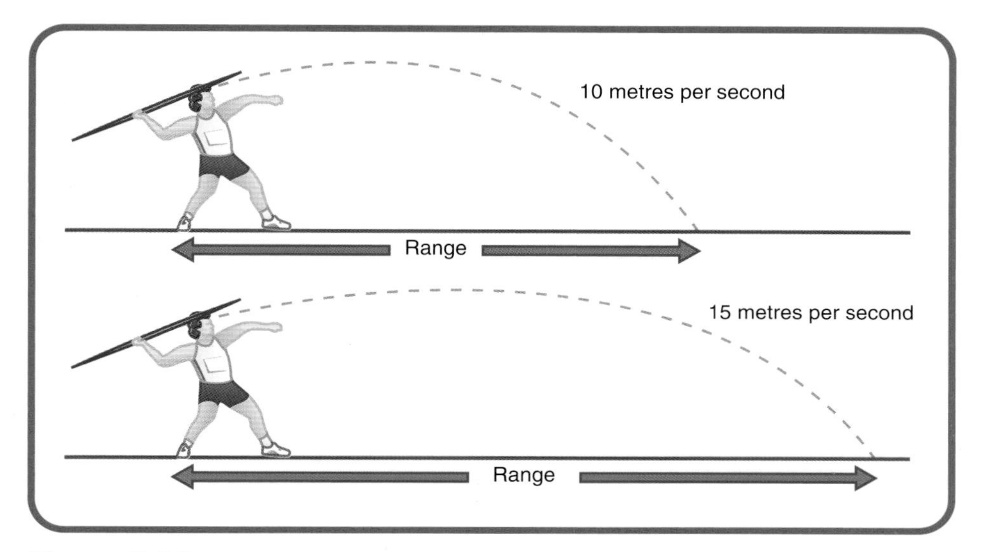


**Figure 5.39**

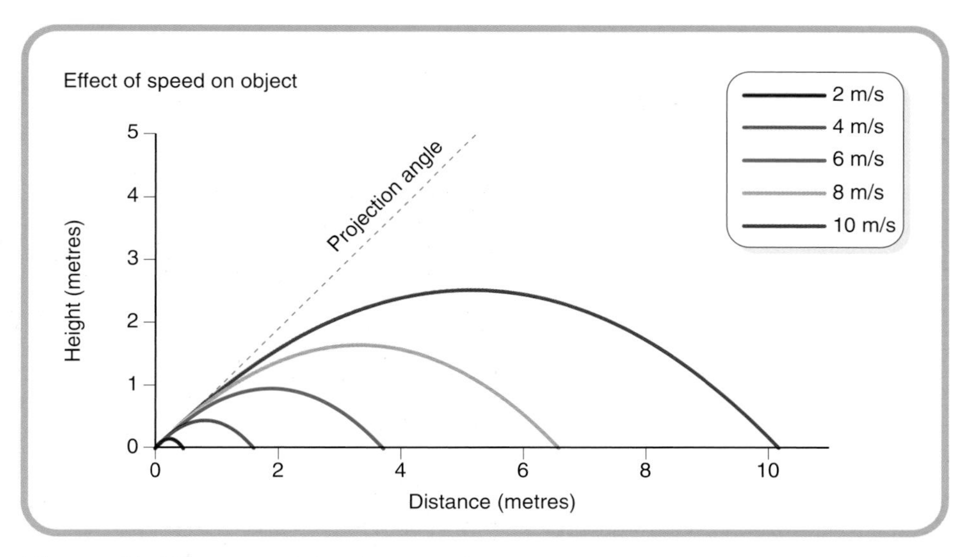


**Figure 5.40**

- In figure 5.40, the greater the speed the greater the range. At a speed of 2 metres per second the range is about 1.5 metres, but at five times the speed, that is 10 metres per second, the range is just over 10 metres.

- The greater the angle the greater the range. In figure 5.41 different ranges are shown for different angles. The speed is kept constant at 10 metres per second. At an angle of 15 degrees the range is 5 metres, but at 45 degrees, the range is over 10 metres.

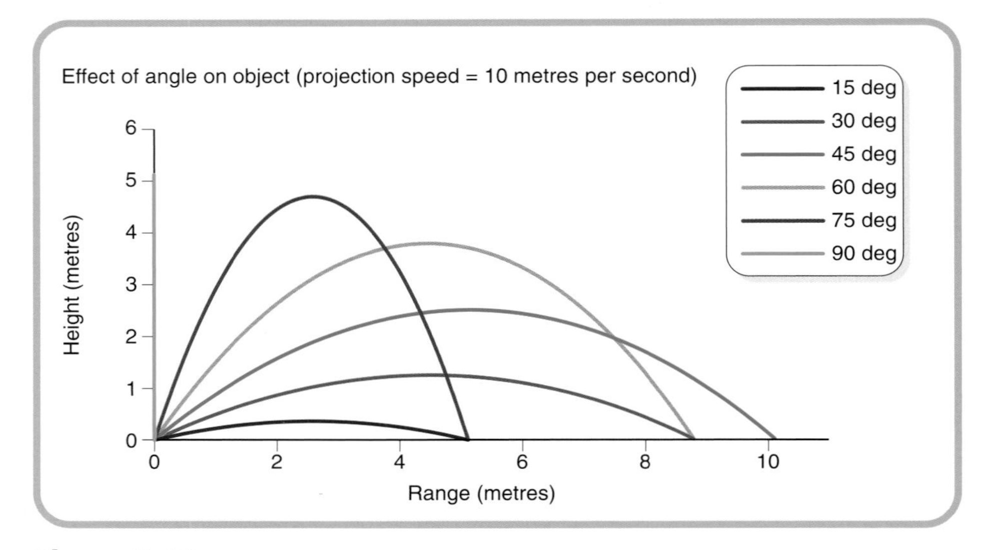


**Figure 5.41**

- In real life it is difficult to obtain the largest angle which is 45 degrees. The body structure of the athlete will be a factor and this may limit both speed and angle. The angles for different sports have been determined and are shown in table 5.3.

| Sport | Actual angle of throwing in degrees |
|---|---|
| Long jump | 18 to 27 |
| Ski jump | 4 to 6 |
| High jump | 40 to 48 |
| Shot put | 36 to 36 |

**Table 5.3**

## Bouncing and height

Bouncing a ball on the ground is used in many sports. Basketball courts are made of a special material which helps the ball bounce (figure 5.42).

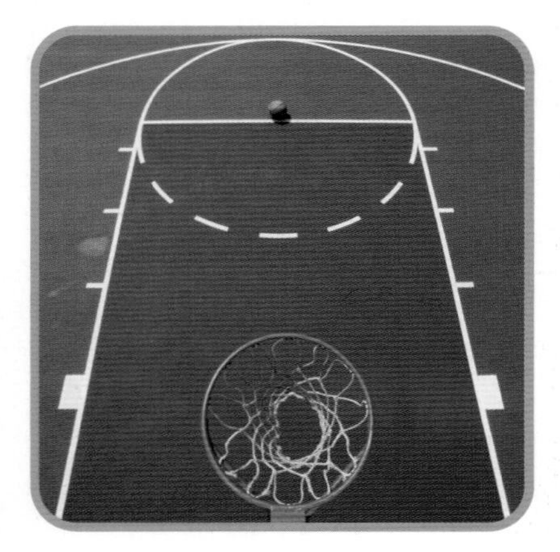

**Figure 5.42**

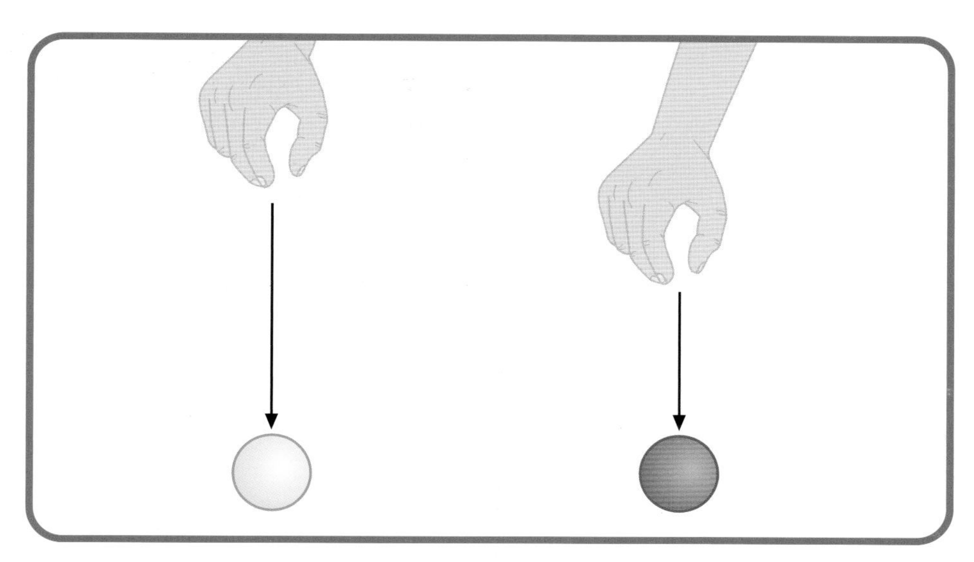

**Figure 5.43**

In figure 5.43 the ball has been bounced from two different heights.

When the ball is dropped from a greater height it will reach the ground at a faster speed.

The faster the speed that the ball hits the ground the greater the height to which it will rebound.

The material that makes up the ground surface also affects the height of rebound.

If the material is very firm like concrete or a tar then the ball will rebound much higher than if a soft surface like grass is used (figure 5.44).

**Figure 5.44** *(a) A grass (soft surface) tennis court* *(b) a clay (hard surface) tennis court.*

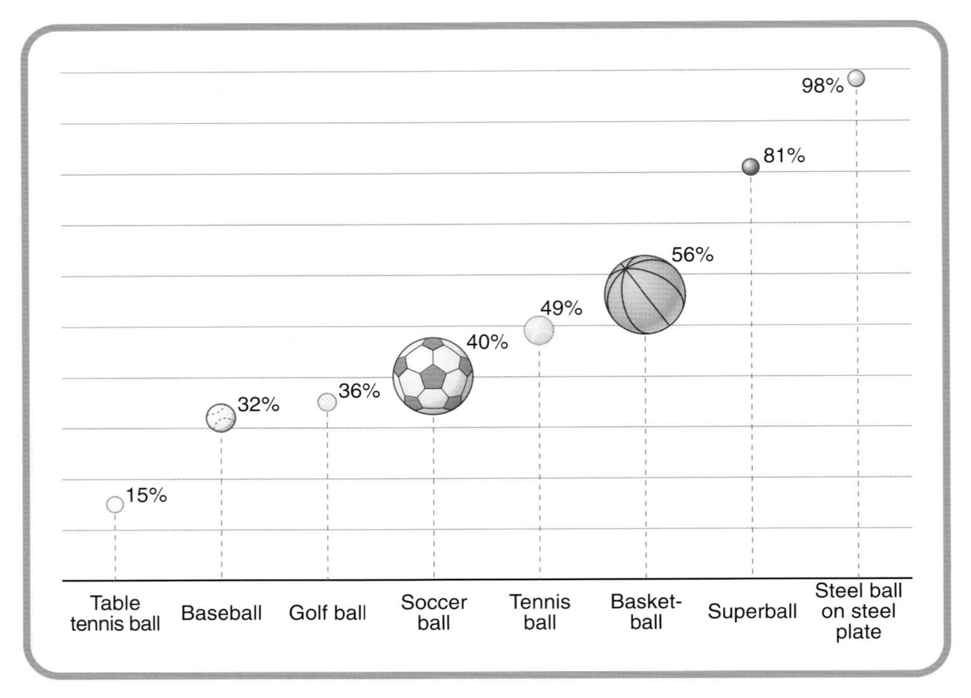


**Figure 5.45**

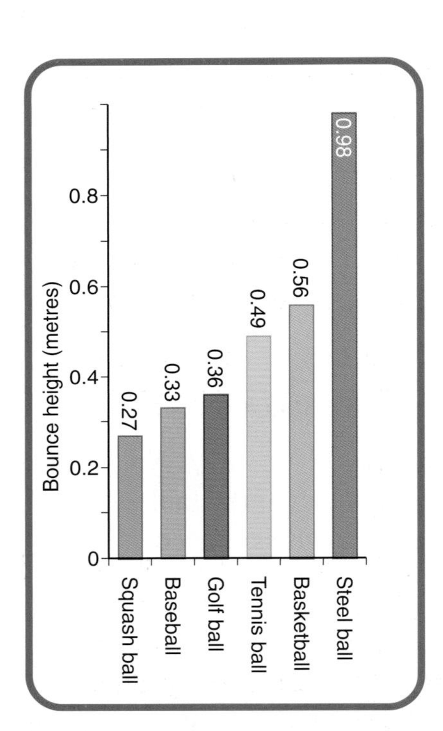


**Figure 5.46**

In the same way a hard ball will rebound to a greater height than a soft ball if dropped from the same height (figure 5.45).

This is shown in the graph of different types of ball when dropped from a height of one metre. The height of the first rebound is measured (figure 5.46).

Why does this happen ?

When a ball is dropped, gravity pulls it towards the floor. The stored energy at the top is changed into kinetic energy. When the ball hits the floor that energy has to be changed into other types of energy. Some of it is used to allow the ball to bounce upwards again. If the ball is very soft like a tennis ball or basketball then as the ball changes shape the energy is stored and then changes back to allow the ball to bounce. If the ball does not bounce very much then most of the energy goes into heat.

## Key Points

- The effect of a collision increases with the mass and speed of the objects involved.
- The change in speed of an object increases with (a) the size of the force acting on it, (b) the time the force acts.
- The range of a ball thrown at an angle is affected by:
  a) the speed of the throw,
  b) the angle of the throw.
- The height that a ball rebounds on hitting a surface is affected by (a) the speed on impact, (b) the material of the surface, (c) the material of the ball.

## Section Questions

**1**

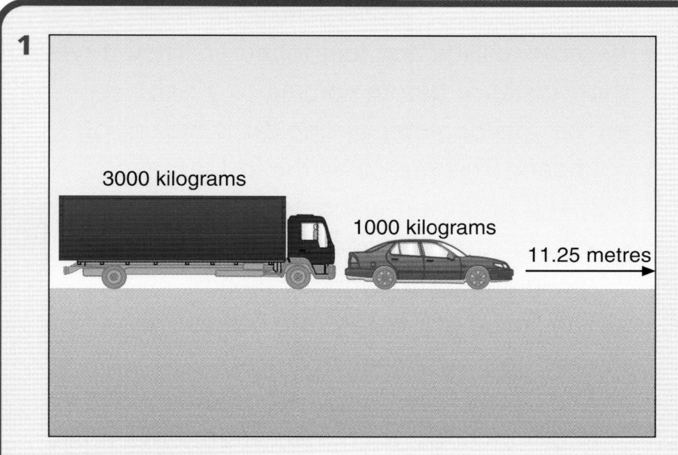


**Figure 5.47**

A lorry of mass 3000 kilograms is shown travelling at 15 metres per second when it strikes a car of mass 1000 kg which is at rest. After the collision they both travel at 11.25 metres per second (figure 5.47). A lorry of the same mass is now travelling at 12 metres per second before a collision with an identical car.

a) Will the speed after the second collision be lower, the same or higher than the first collision?

b) A lorry of mass 4500 kg is travelling at 15 metres per second. The lorry collides with an identical car as before. What will happen to the speed after the collision?

**2** A tennis player has to have his racquet restrung. If he wants to make the ball travel further when struck, what should happen to the strings?

a) strung as slack as possible,

b) strung as tight as possible.

**3** In the early days of golf, balls were made of soft materials inside a coating. Modern balls are made of a harder material.

a) When struck in exactly the same way, which ball will travel further?

b) Explain your answer

**4** A golf ball is struck and hits a bunker made of sand. When an identical ball is struck at the same speed and angle it hits a grass fairway.

a) Which one will continue to bounce after striking the surface?

b) Explain your answer

**5** An archer fires an arrow from her bow. To increase the distance travelled by the arrow, the archer pulls the string of the bow back further than before.

a) What change will this make to the speed of the arrow as it leaves the bow?

b) What other change could the archer make to increase the distance travelled by the arrow?

**6** An egg is dropped onto a concrete surface. The experiment is repeated with an egg dropped onto a foam sponge.

a) Which egg will smash?

b) Explain your answer.

**7** When a tennis ball is dropped from 1.5 metres it rebounds to a height of 1 metre. A golf ball dropped from the same height rebounds to a height of 1.2 metres. What can you say about the material the golf ball is made from compared to the tennis ball?

**8** To test tennis balls, the balls are dropped onto a clay surface. They are dropped from a height of 2 metres and the height of the rebound measured. The experiment is repeated on soft grass. What difference would you expect in the results of the two experiments?

## Examination Style Questions

**1** Some information about cars is shown in table 5.4.

| Car | Maximum speed in miles per hour | Time in seconds to go from 0 to 60 miles per hour |
|---|---|---|
| Wellington | 102 | 13.2 |
| Xciting | 105 | 11.4 |
| Ying | 104 | 12.5 |
| Zeus | 106 | 11.6 |

**Table 5.4**

a) Which car has the greatest maximum speed?
b) Which car has the largest acceleration?
c) When the Ying car reaches its maximum speed, it cannot go any faster even if the accelerator pedal is pushed further down.
   i) What forces are acting on the car?
   ii) Are the forces balanced of unbalanced?
d) Describe two ways in which a car could be streamlined.

**2** Two students want to measure the average speed of a car as it travels along a straight stretch of road.
a) What apparatus will they need?
b) What measurements should they make?
c) How would they use these measurements to calculate the average speed of the car?
d) The road distance is measured to be 32 metres and the time for a car to travel this distance is 4 seconds. Calculate the average speed of the car.

**3** Two cars collide and join together. They travel a short distance before coming to a halt.
a) The speedometer of one car is broken on impact. This measures the instantaneous speed of the car just before the collision. What is the difference between average speed and instantaneous speed?
b) The speed just before the collision is measured by a speed camera.
The car travelled between the marks shown in 0.5 seconds. Each mark is 2 metres apart. Calculate the speed of the car (figure 5.48).
c) What **two** changes to the car would have made the force of the collision greater?
d) The mass of one car is 1200 kg. What is its weight on Earth?

**Figure 5.48**

**4** A student is trying to improve her golf. She wants to increase the distance she can drive the ball from the tee.
a) What two changes could she make to increase the range of the ball?
b) She is told to alter her golf swing so that she hits the ball with greater force. What will this do to the range the ball travels?
c) The ball goes further along the ground on a dry day than a wet day. Explain why this happens in terms of the force of friction.

# Electronics

**The electronics industry has brought us an amazing number of electronic gadgets in recent years – compact disc music systems and games consoles, for example. This chapter breaks down some complex electronic systems into simpler parts, which can be more easily understood. It also looks at how digital logic circuits can be used to solve everyday problems.**

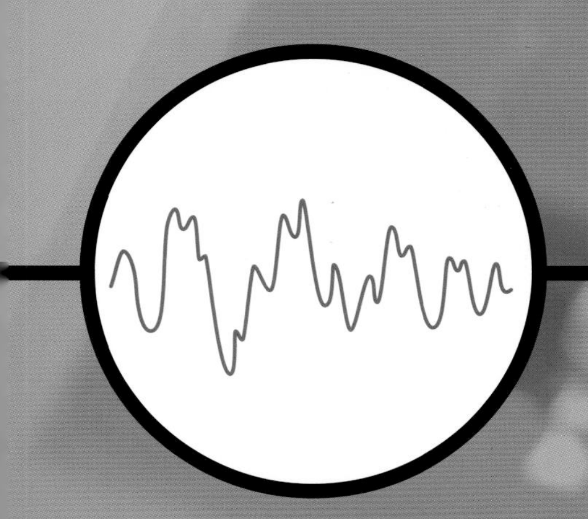

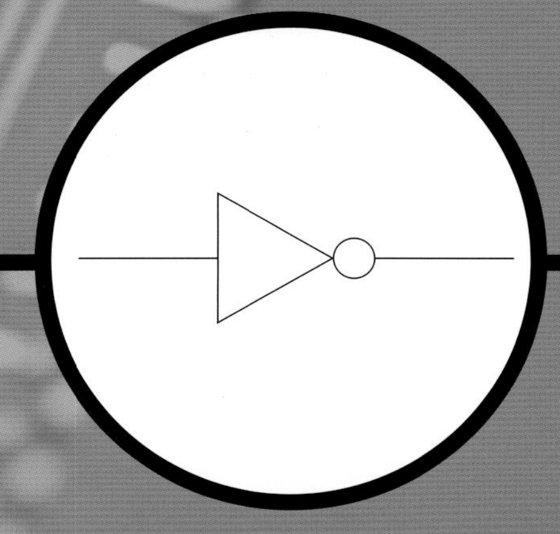

# 1 Input, process and output

## After studying this section you should be able to...

1 State that an electronic system consists of three parts: input, process and output.
2 Identify from a block diagram the input, process and output sub-systems of an electronic system.
3 Draw a block diagram showing the input, process and output sub-systems of an electronic system.
4 State that the microphone, thermistor, LDR and switch are examples of input devices.
5 State that a microphone changes sound energy to electrical energy.
6 State that the resistance of a thermistor changes with temperature.
7 State that the resistance of an LDR decreases as the light gets brighter.
8 Identify from a list an appropriate input device for a given application.
9 State that an output device changes electrical energy into another form of energy.
10 State that the loudspeaker, buzzer, lamp, LED and electric motor are examples of output devices.
11 State the energy transformations involved for a given output device.
12 Identify from a list an appropriate output device for a given application.

**Figure 6.1** *A baby alarm.*

An **electrical system** is a collection of components connected together to perform a particular job, for example a baby alarm (figure 6.1).

A baby alarm consists of three main parts or sub-systems:

- input — a microphone
- process — an amplifier
- output — a loudspeaker.

All electronic systems can be broken into these three main parts or sub-systems — **input, process and output**. The input sub-system starts the system working by detecting some type of energy and changing it into electrical energy.

The process sub-system alters the electrical energy from the input so as to produce the required electrical energy needed to work the output.

The output sub-system converts the electrical energy from the process into another type of energy so that the job can be done.

An electronic system is often drawn as a **block diagram** (figure 6.2). The arrows show how information is passed (electrically) from one block to another.

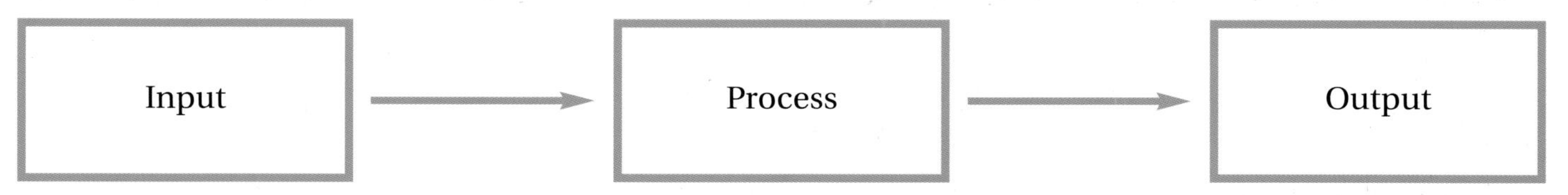


**Figure 6.2**

The block diagrams for a number of electrical systems are shown in figure 6.3.

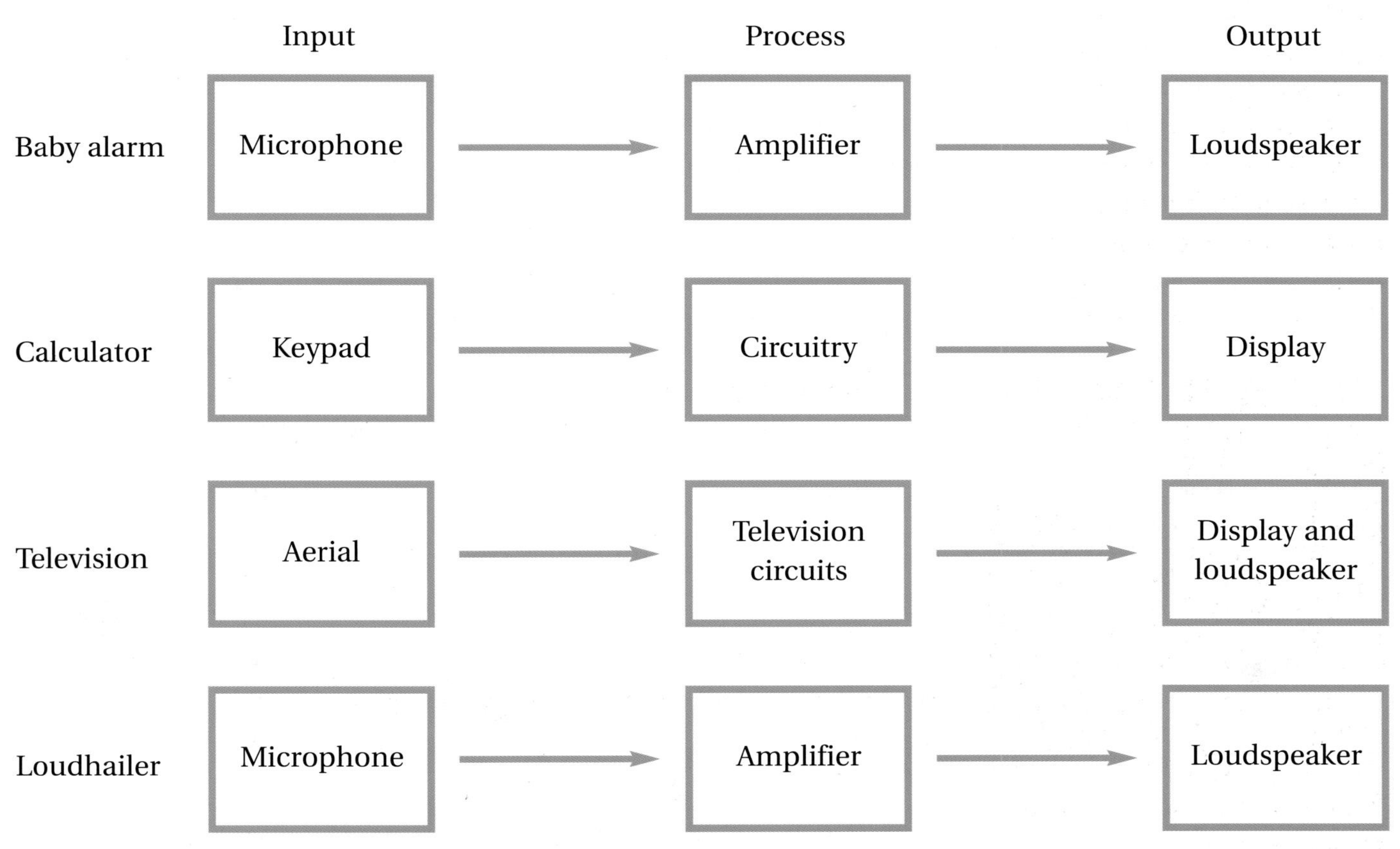


**Figure 6.3**

## Input devices

### Microphone

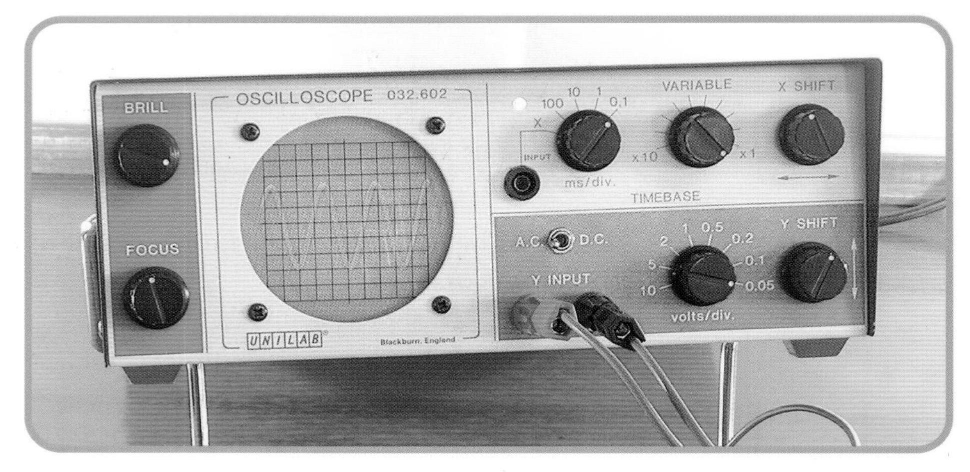


**Figure 6.4** *An oscilloscope.*

Connect a microphone to an oscilloscope (figure 6.4). Whistle a quiet note and then a louder note into the microphone. Look at the trace obtained on the oscilloscope screen in each case. What is the energy change that takes place at the microphone?

A microphone is connected to an oscilloscope. As louder notes are played into the microphone, the trace on the oscilloscope screen increases in size.

A **microphone** is an input device which changes sound energy into electrical energy. The louder the sound, the greater the electrical energy produced.

### Thermistor

**Figure 6.5** *A thermistor connected to an ohmmeter.*

Connect a thermistor to an ohmmeter (figure 6.5). Place the thermistor in a beaker of (a) cold water and then (b) warm water. Look at the reading on the ohmmeter in each case.

When a thermistor is heated the ohmmeter reading decreases.

A **thermistor** is an input device. The resistance of a thermistor changes when its temperature changes.

### Light-dependent resistor (LDR)

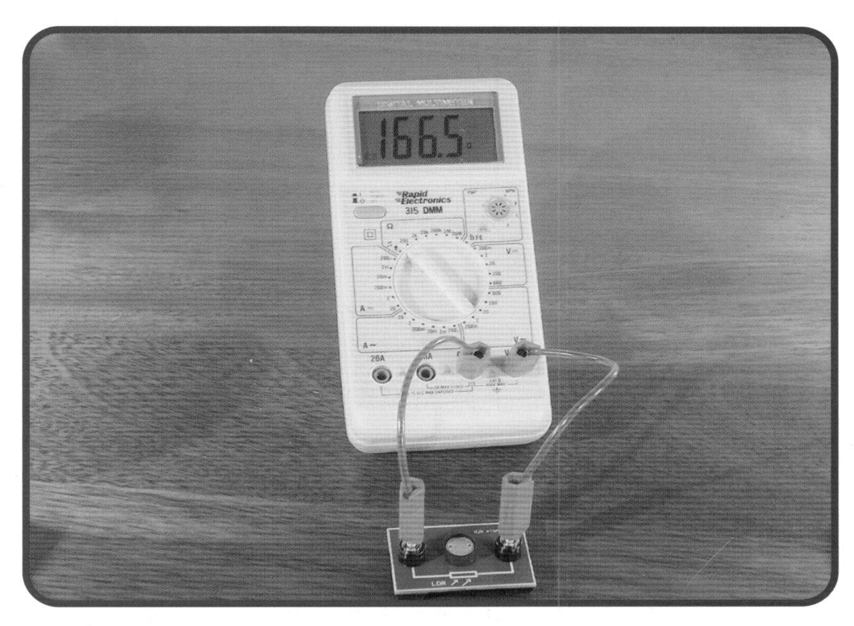


**Figure 6.6** *A light dependent resistor connected to an ohmmeter.*

Connect a light-dependent resistor (LDR) to an ohmmeter (figure 6.6). Cover the LDR with your hand and then place it in bright light. Look at the reading on the ohmmeter in each case.

When the light-dependent resistor is exposed to more light, the ohmmeter reading decreases.

A light-dependent resistor (LDR) is an input device. As the light gets brighter (light intensity increases) the resistance of the LDR decreases.

### Switch

**Figure 6.7** *A switch connected to an ohmmeter.*

Connect a switch to an ohmmeter (figure 6.7). Look at the readings on the ohmmeter when the switch is open and closed.

An ohmmeter is connected to a switch. When the switch is open, there is a gap between the contacts of the switch and the reading displayed by the ohmmeter is very, very high – the resistance of an open switch is very, very big.

When the switch is closed, the contacts touch and the reading displayed by the ohmmeter is zero (or very close to zero).

### Summary of input devices

- Microphone — changes sound energy into electrical energy.
- Details of input devices are given in table 6.1.

| Device | Conditions | Resistance |
|---|---|---|
| Thermistor | Low temperature | High |
| | High temperature | Low |
| LDR | Dark | High |
| | Light | Low |
| Switch | Open | Very, very high |
| | Closed | Zero |

**Table 6.1**

## Output devices

### Loudspeaker

Connect a loudspeaker to a signal generator. Change the amplitude (loudness) and frequency controls on the signal generator. What is the energy change that takes place at the loudspeaker? Name some appliances in your home which contain a loudspeaker.

A loudspeaker is an output device which changes electrical energy into sound energy. A radio and a television are examples of electronic systems that contain a loudspeaker.

### Buzzer

Connect a buzzer to a battery and a switch. Open and close the switch. What is the energy change that takes place at the buzzer? Name an appliance which could contain a buzzer.

A buzzer is an output device which changes electrical energy into sound energy. A microwave is an example of an electronic system that often contains a buzzer.

### Electric motor

Connect an electric motor to a battery and a switch. Open and close the switch. What is the energy change that takes place at the electric motor? Name some appliances which contain an electric motor.

An electric motor is an output device which changes electrical energy into kinetic (movement) energy. Vacuum cleaners and washing machines contain mains-operated electric motors.

### Lamp

Connect a lamp to a battery and a switch. Open and close the switch. What is the energy change that takes place in the lamp?

A lamp consists of a thin wire (filament) in a glass container. When an electric current passes through the wire, electrical energy is changed into light and heat in the filament. Mains-operated lamps are used to light our homes.

### Light-emitting diode (LED)

Connect a light-emitting diode (LED) to a battery and a switch. Open and close the switch. What happens?

Reverse the connections to the diode. Open and close the switch. What happens?

What is the energy change that takes place at the LED? Name an appliance which contains an LED.

When an LED is correctly connected to a d.c. supply an electric current passes through the LED and the LED emits light. The LED does not light up if the connections to the d.c. power supply are then reversed. LEDs are available in red, green, yellow, blue and white colours. LEDs can be used in hi-fi equipment and instrument panels.

## Key Points

- An electrical system consists of three parts – input, process and output.
- Most input devices change some form of energy into electrical energy.
- Examples of input devices are a microphone, thermistor, LDR and a switch.
- A microphone changes sound energy into electrical energy.
- The resistance of a thermistor changes with temperature.
- An open switch has a very, very high (infinite) resistance and a closed switch has zero resistance.
- The resistance of an LDR decreases as the light gets brighter.
- An output device changes electrical energy into some other form of energy.
- Examples of output devices are a loudspeaker, buzzer, lamp, LED and electric motor.
- A loudspeaker and a buzzer change electrical energy into sound energy.
- A lamp and an LED change electrical energy into light energy.
- An electric motor changes electrical energy into kinetic (movement) energy.

## Section Questions

**1** Electronic systems consist of three parts. Name these three parts.

**2** The block diagram for a radio is shown in figure 6.8.

**Figure 6.8**

Which of these three blocks represents the process device for the radio?

**3** A calculator can be broken down into three sub-systems. These sub-systems are the circuit, the display and the keypad.

Copy and complete the sub-systems in the block diagram for the calculator in figure 6.9.

**Figure 6.9**

**4** Four input devices are listed below.

microphone light dependent resistor (LDR)
switch thermistor

From the above list, choose an appropriate input device which could form part of the input sub-system:
a) for an electronic light meter,
b) to measure the temperature of an oven,
c) for a public address system,
d) for a contestant to press in a quiz game,
e) an electronic sound sensor.

**5** A light level meter has a light dependent resistor (LDR) as part of its input sub-system. What happens to the resistance of the LDR when the light meter is moved to a more brightly lit room?

**6** Five output devices are listed below.

buzzer lamp LED
loudspeaker electric motor

From the above list, choose an appropriate output device which could form part of the output sub-system to:
a) move a conveyor belt at a supermarket check-out,
b) give a visual indication to a car driver that a door is open,
c) indicate that a microwave programme is complete.

**7** State the energy conversion for the following output devices:
a) buzzer, b) electric motor, c) lamp, d) LED, e) loudspeaker.

**8**

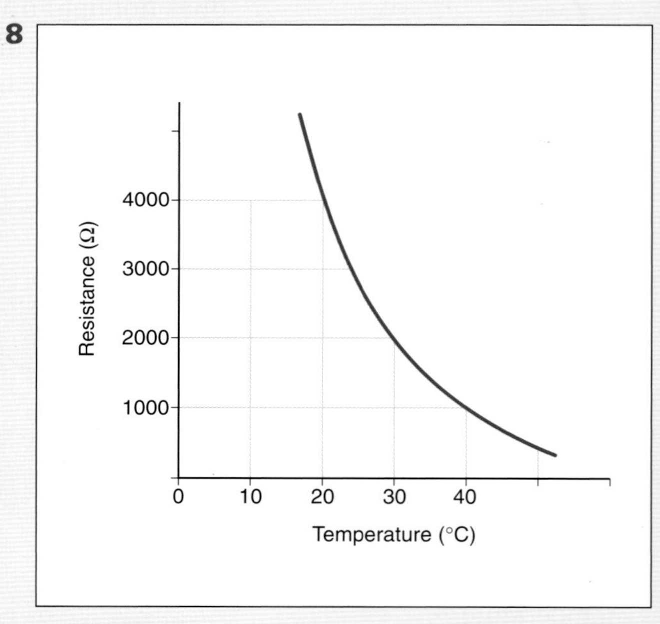


**Figure 6.10**

A thermistor is used as a temperature sensor. The graph in figure 6.10 shows how the resistance of the thermistor changes as the temperature changes.
a) What is the resistance of the thermistor at 20 °C?

b) The current in the thermistor at 20 °C is 0.0050 amperes. The temperature of the thermistor rises to 22 °C.
   (i) Which of the following is most likely to be the current in the thermistor at 25 °C.
      (A) 0.0040 amperes
      (B) 0.0050 amperes
      (C) 0.0060 amperes.
   (ii) Explain your answer.

**9** Some electronic components are listed below.

buzzer    electric motor    LDR
LED    microphone    thermistor

Copy and complete table 6.2 to show which of these components are input devices and which are output devices.

| Input devices | Output devices |
|---|---|
| | |
| | |
| | |
| | |

**Table 6.2**

# Digital logic gates

## After studying this section you should be able to...

1. Draw and identify the symbols for two input AND and OR gates, and a NOT gate.
2. State that high voltage = logic 1 and low voltage = logic 0.
3. State that for a NOT gate the output is the opposite of the input.
4. State that for an AND gate both inputs must be high for the output to be high.
5. State that for an OR gate either input must be high for the output to be high.
6. Explain how to use combinations of digital logic gates for control in simple situations.

## Analogue signals and digital signals

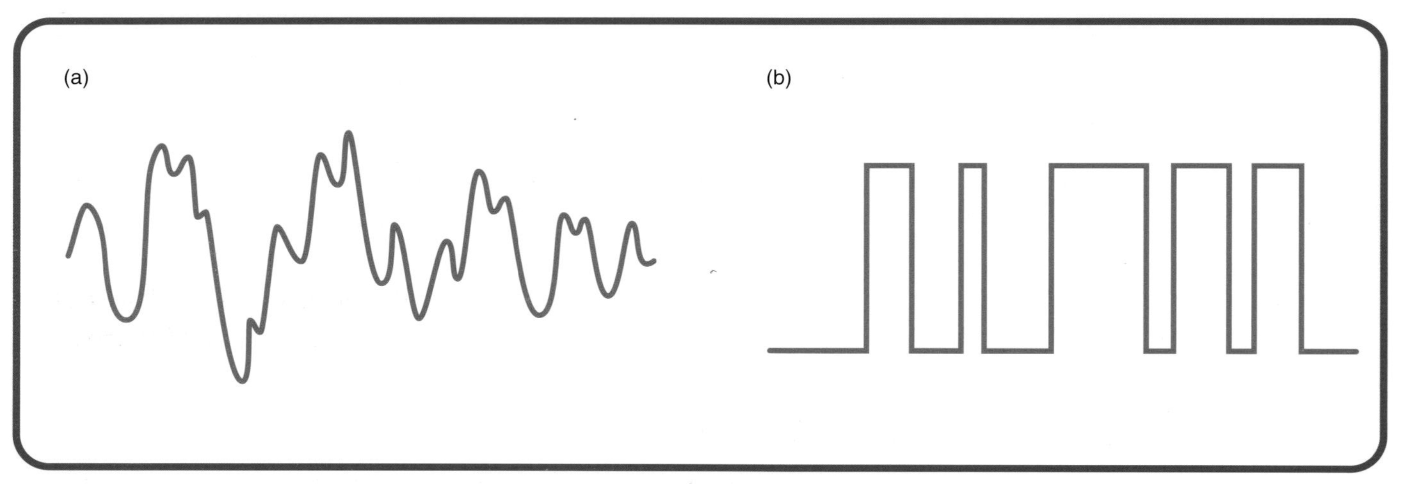


**Figure 6.11** *(a) An analogue signal, (b) a digital signal.*

The signals used by electronic systems are of two types: **analogue** or **digital**. Figure 6.11 shows the traces displayed on an oscilloscope screen of (a) an analogue signal and (b) a digital signal.

Figure 6.11(a) shows a typical electrical signal from the microphone of a telephone when a person is speaking. The trace has a continuous range of values. This type of signal is called an analogue signal.

Figure 6.11(b) shows a typical electrical signal from a compact disc player. The trace has a series of electrical pulses each with the same height (amplitude). This type of signal is called a digital signal.

An analogue signal has a continuous range of values, while a digital signal can have only one of two possible values.

## Logic gates

Logic gates are digital devices. The inputs and outputs from a logic gate are voltages which may either be on (high voltage) or off (low voltage). For logic gates a high voltage = logic 1 and low voltage = logic 0.

A table known as a truth table shows how the output of the gate varies with the input or inputs.

## The NOT-gate

Figure 6.12 shows the circuit symbol and truth table for a **NOT-gate**.

| Input | Output |
|---|---|
| 0 | 1 |
| 1 | 0 |

**Figure 6.12** *(a) The circuit symbol and (b) the truth table for a NOT-gate.*

From the truth table it can be seen that the output of a NOT-gate is 'not' (the same as) the input. A NOT-gate is also called an **inverter**.

## The AND-gate

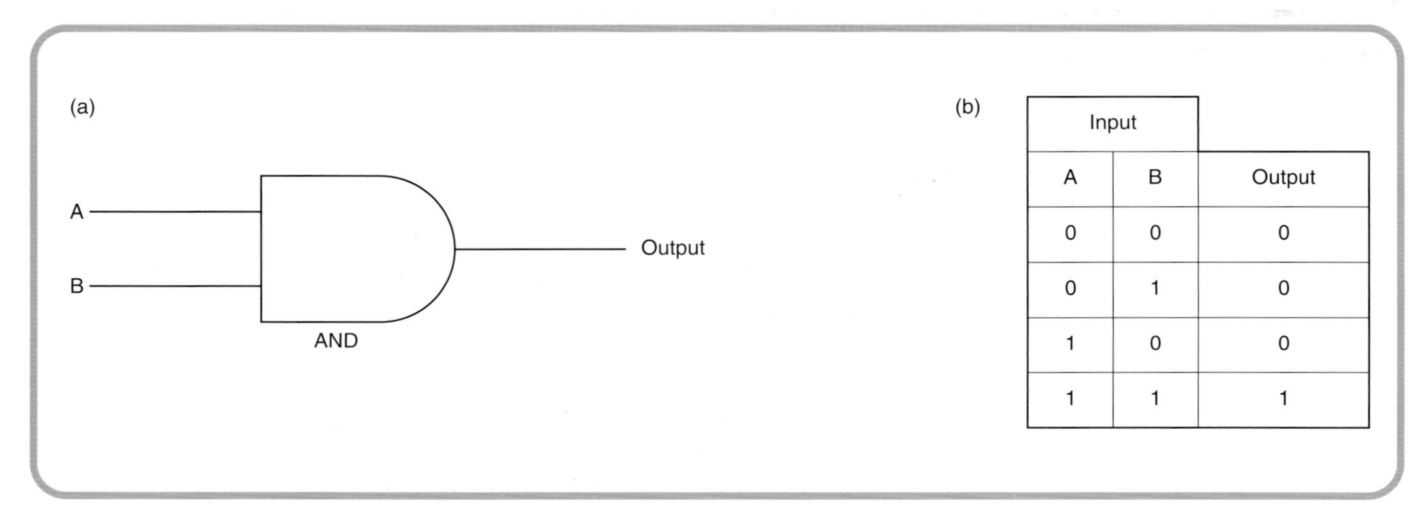


| Input | | |
|---|---|---|
| A | B | Output |
| 0 | 0 | 0 |
| 0 | 1 | 0 |
| 1 | 0 | 0 |
| 1 | 1 | 1 |

**Figure 6.13** *(a) The circuit symbol and (b) the truth table for an AND-gate.*

The circuit symbol and truth table for an **AND-gate** are shown in figure 6.13.

From the truth table it can be seen that the output of an AND-gate will be logic 1 (high) only when inputs A and B are both logic 1 (high).

## The OR-gate

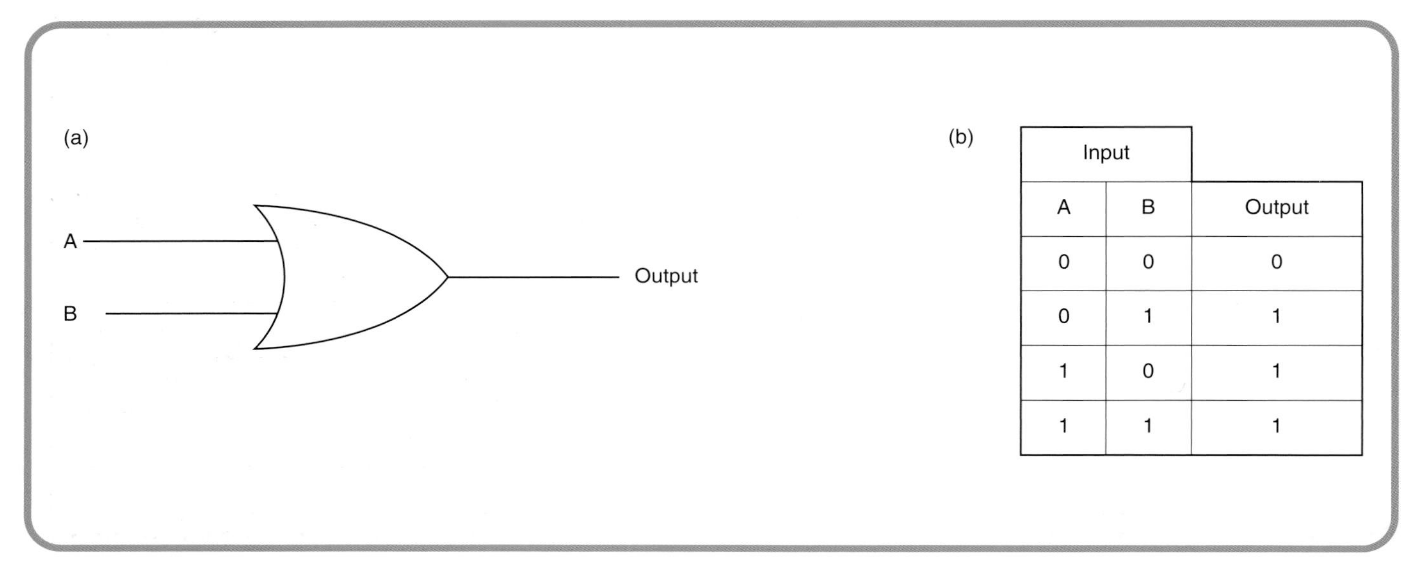


| Input | | |
|---|---|---|
| A | B | Output |
| 0 | 0 | 0 |
| 0 | 1 | 1 |
| 1 | 0 | 1 |
| 1 | 1 | 1 |

**Figure 6.14** *(a) The circuit symbol and (b) the truth table for an OR-gate.*

The circuit symbol and truth table for an **OR-gate** are shown in figure 6.14.

From the truth table it can be seen that the output of an OR-gate will be logic 1 (high) when either of the inputs A or B is logic 1 (high).

In the following examples you may assume that:

- a light sensor gives out a logic 1 (high) in light and a logic 0 (low) in dark,
- a temperature sensor gives out a logic 1 (high) when warm and a logic 0 (low) when cold,
- a switch gives out a logic 1 (high) when closed and a logic 0 (low) when open.

## Examples

Draw a logic diagram for a warning LED to light when a car engine gets too hot. The LED should only operate when the ignition of the car is switched on (logic 1).

Solution
Require LED to be on (1) when the ignition is on (1) and engine is too hot (1) (figure 6.15).

**Figure 6.15**

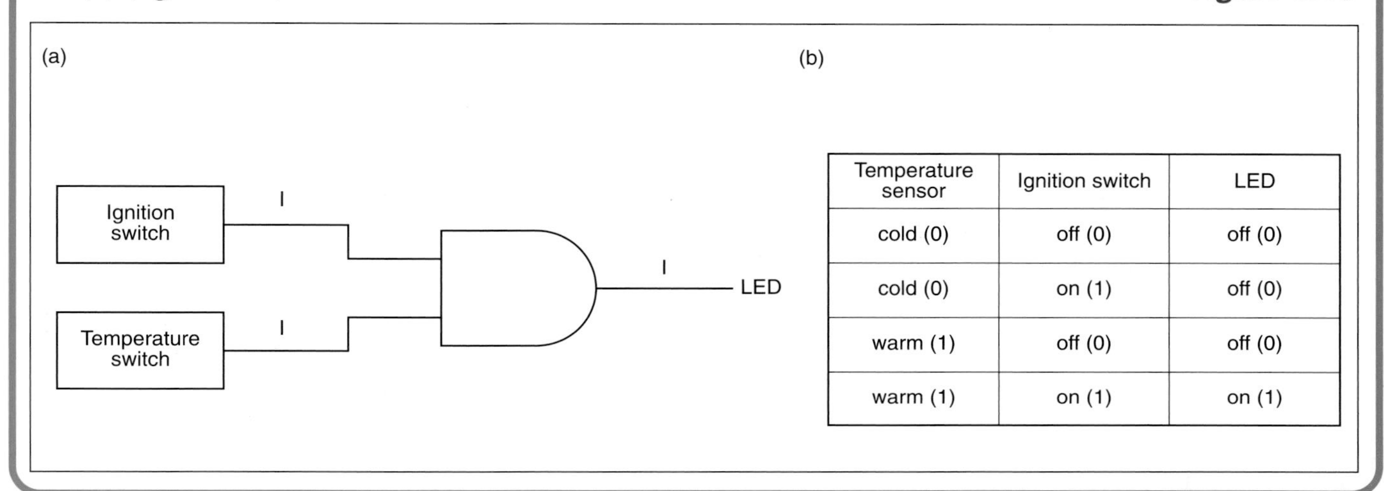


| Temperature sensor | Ignition switch | LED |
|---|---|---|
| cold (0) | off (0) | off (0) |
| cold (0) | on (1) | off (0) |
| warm (1) | off (0) | off (0) |
| warm (1) | on (1) | on (1) |

## Examples

Draw a logic diagram which will switch on the pump of a central heating system when the house is cold and the central heating is switched on (logic 1).

Solution
Require pump on (1) when central heating is on (1) and temperature is cold (0) (not warm (1)) (figure 6.16).

**Figure 6.16**

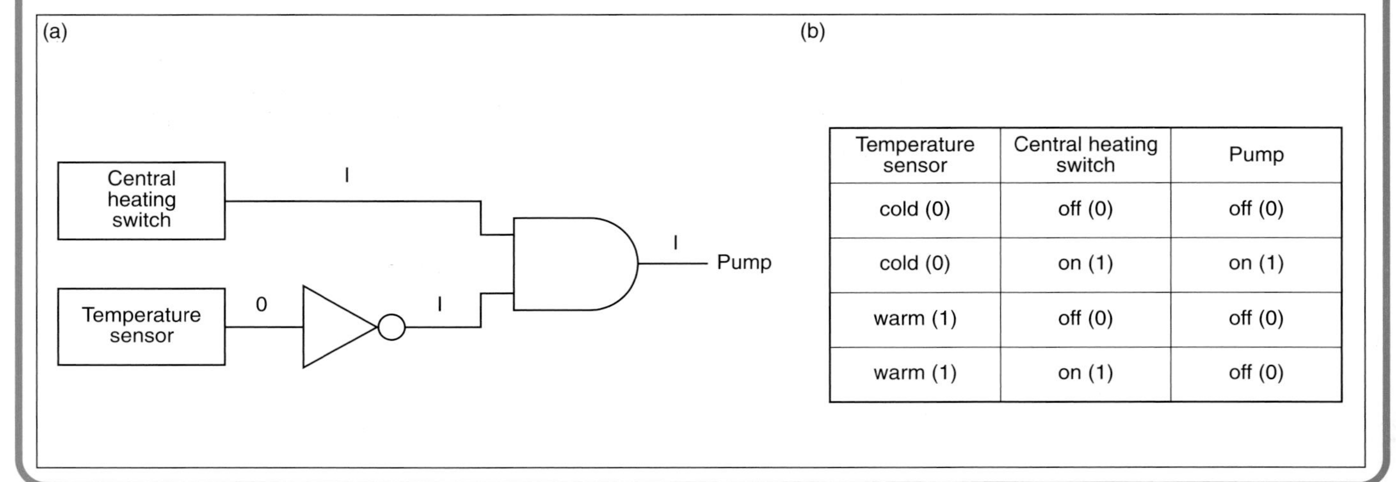


| Temperature sensor | Central heating switch | Pump |
|---|---|---|
| cold (0) | off (0) | off (0) |
| cold (0) | on (1) | on (1) |
| warm (1) | off (0) | off (0) |
| warm (1) | on (1) | off (0) |

## Examples

Draw a logic diagram which will turn on a heater in a greenhouse when it gets cold at night. The heater should be switched off (logic 0) during the day.

Solution
Require heater on (1) when it is cold (0) (not warm (1)) and dark (0) (not light (1)) (figure 6.17).

**Figure 6.17**

(a)

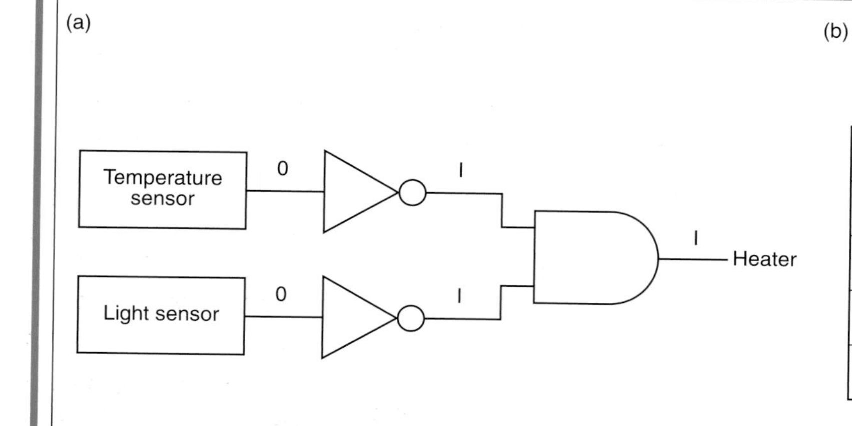


(b)

| Light sensor | Temperature sensor | Heater |
|---|---|---|
| dark (0) | cold (0) | on (1) |
| dark (0) | warm (1) | off (0) |
| light (1) | cold (0) | off (0) |
| light (1) | warm (1) | off (0) |

## Key Points

- Logic gates are digital devices – NOT (or inverter), AND and OR gates.
- Logic gates use logic 1 to represent a high voltage level and logic 0 to represent a low voltage level.
- The circuit symbols and truth tables for NOT, AND and OR gates are shown in figures 6.12, 6.13 and 6.14.

## Section Questions

1

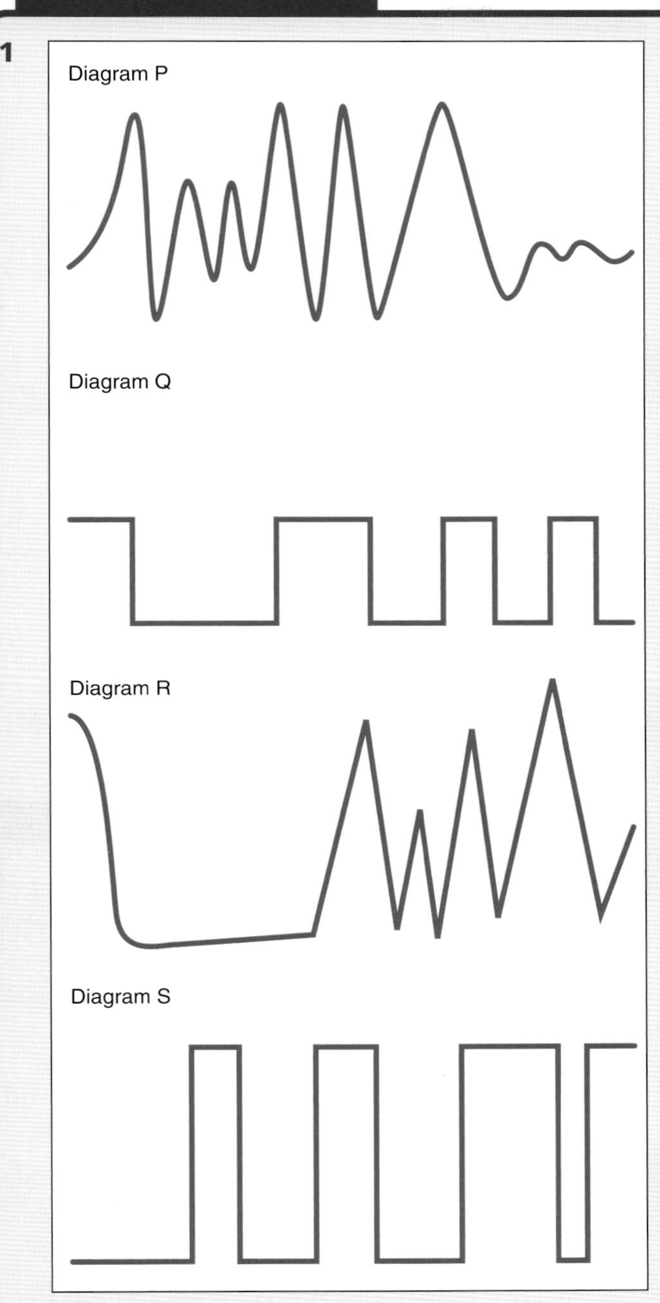


**Figure 6.18**

The output signals from four devices are displayed on the oscilloscope screens shown in figure 6.18.

a) Which of the oscilloscope traces show a digital signal?

b) Which of the oscilloscope traces show an analogue signal?

2 The terms high voltage and low voltage are used when using logic gates. Which of these terms refer to logic 1 and which to logic 0?

3 a) Draw the circuit symbol for a NOT gate.

b) Table 6.3 shows the possible inputs for a NOT gate. Copy and complete the table to show the correct outputs.

| Input | Output |
|---|---|
| 0 | |
| 1 | |

**Table 6.3**

4 a) Draw the circuit symbol for an AND gate.

b) Table 6.4 shows the possible inputs for an AND gate. Copy and complete the table to show the correct outputs.

| Input A | Input B | Output |
|---|---|---|
| 0 | 0 | |
| 0 | 1 | |
| 1 | 0 | |
| 1 | 1 | |

**Table 6.4**

5 a) Draw the circuit symbol for an OR gate.

b) Table 6.5 shows the possible inputs for an OR gate. Copy and complete the table to show the correct outputs.

| Input A | Input B | Output |
|---|---|---|
| 0 | 0 | |
| 0 | 1 | |
| 1 | 0 | |
| 1 | 1 | |

**Table 6.5**

6 State an alternative name for the logic gate known as an inverter.

**7**

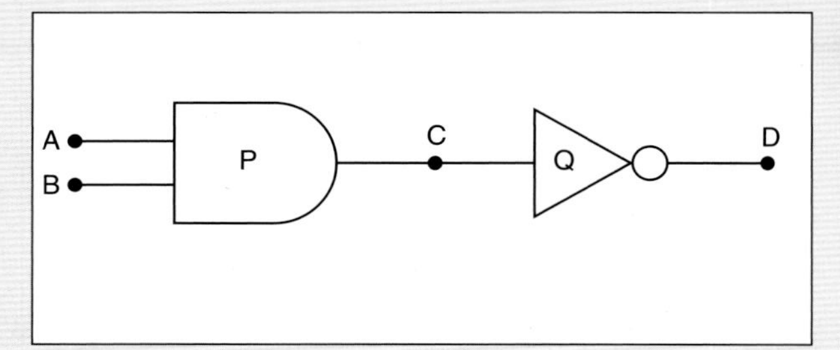


**Figure 6.19**

Part of an electronic circuit is shown in figure 6.19.

a) What name is given to:
   (i) gate P?
   (ii) gate Q?

b) Table 6.6 shows the possible logic levels of inputs A and B. Copy and complete the table to show the logic levels of C and D.

| A | B | C | D |
|---|---|---|---|
| 0 | 0 | | |
| 0 | 1 | | |
| 1 | 0 | | |
| 1 | 1 | | |

**Table 6.6**

**8**

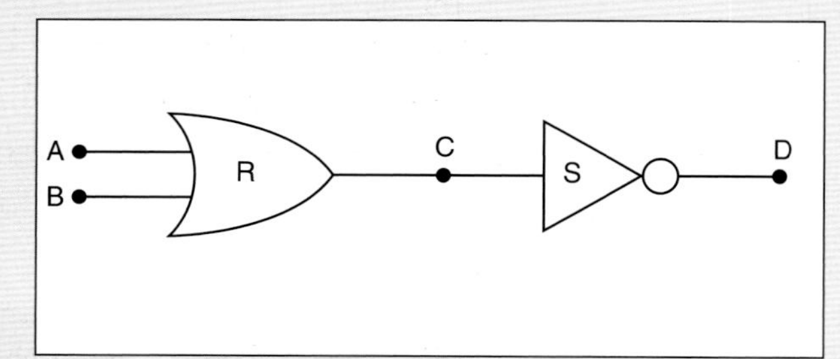


**Figure 6.20**

Part of an electronic circuit is shown in figure 6.20.

a) What name is given to:
   (i) gate R?
   (ii) gate S?

b) Table 6.7 shows the possible logic levels of inputs A and B. Copy and complete the table to show the logic levels of C and D.

| A | B | C | D |
|---|---|---|---|
| 0 | 0 | | |
| 0 | 1 | | |
| 1 | 0 | | |
| 1 | 1 | | |

**Table 6.7**

**9**

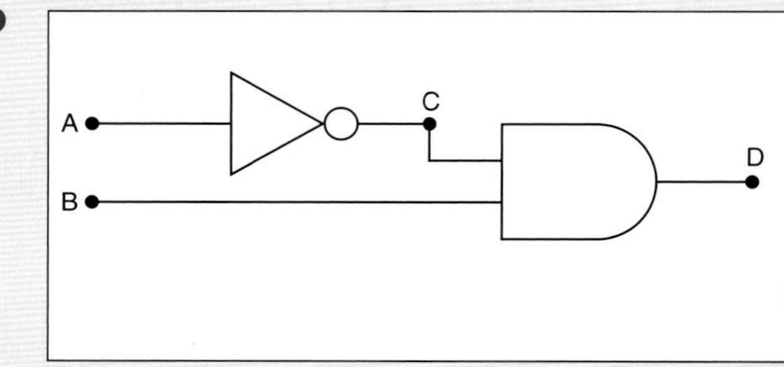


**Figure 6.21**

Table 6.8 shows the possible logic levels of inputs A and B (see figure 6.21). Copy and complete this table to show the logic levels of C and D.

| A | B | C | D |
|---|---|---|---|
| 0 | 0 | | |
| 0 | 1 | | |
| 1 | 0 | | |
| 1 | 1 | | |

**Table 6.8**

**10** A student is asked to design a circuit to remind a car driver that the car lights have been left on after the ignition switch has been switched off. A buzzer is to sound when the lights are on but only when the ignition is switched off.
The light switch sensor gives logic 1 when the lights are on.
The ignition switch gives logic 1 when switched on.
The buzzer is switched on by logic 1.
Draw a suitable logic diagram, naming the logic gates involved.

## Examination Style Questions

**1** A shop sells the following electronic components:
buzzer electric motor light dependent resistor (LDR) light emitting diode (LED) microphone thermistor

a) (i) From the above list, name two output devices.
(ii) From the above list, name two input devices.
(iii) State the useful energy change that takes place in a microphone.

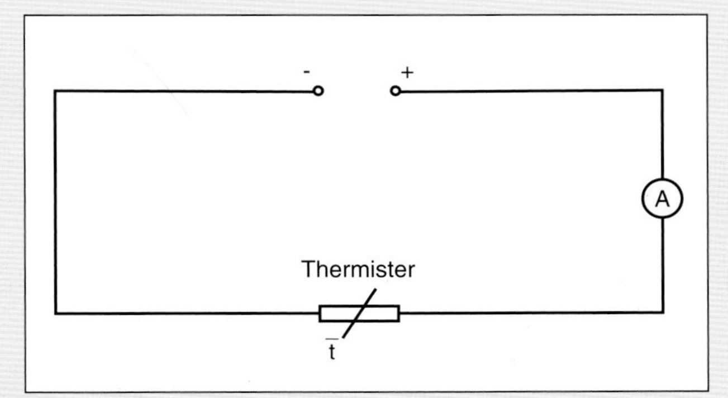


**Figure 6.22**

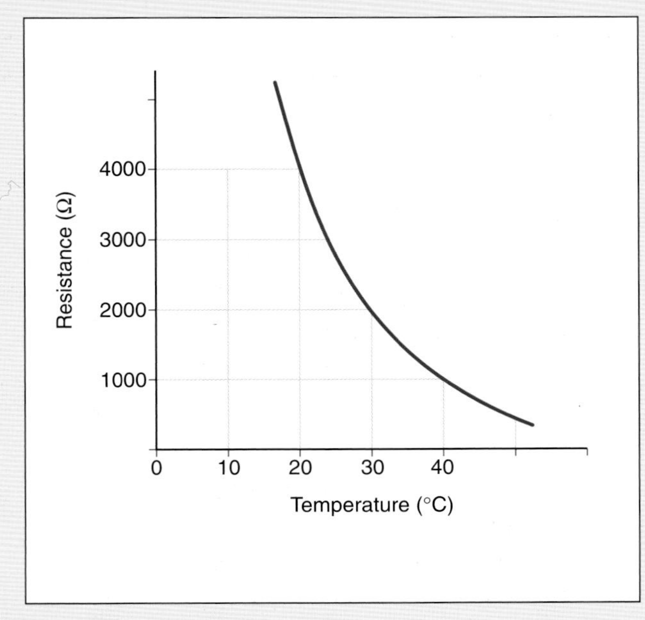


**Figure 6.23**

b) A thermistor is connected in the circuit shown in figure 6.22.

The graph in figure 6.23 shows how the resistance of the thermistor changes as the temperature changes.
During the day the temperature of the thermistor rises.
(i) What happens to the reading on the ammeter?
(ii) Explain your answer.

**2** a) An electronic system is represented by the block diagram shown in figure 6.24.

**Figure 6.24**

Copy figure 6.24 and insert the missing labels to complete the block diagram.
b) Name an input device which changes sound energy to electrical energy.
c) Name an input device which changes light energy to electrical energy.

**3** The logic diagram for the operation of a car interior lamp is shown in figure 6.25.

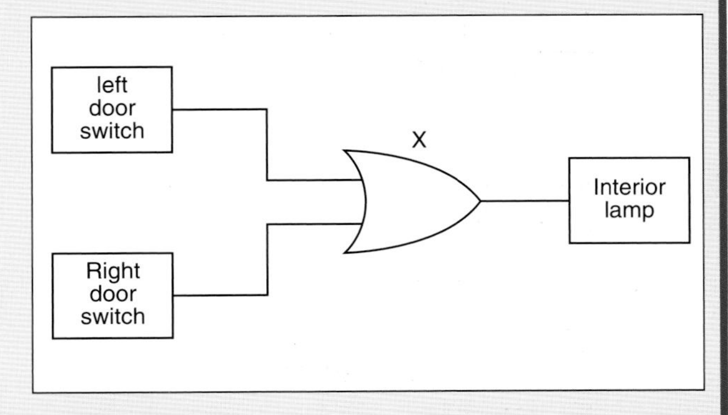


**Figure 6.25**

The interior lamp will come on when either door is open.
When the output from gate X is logic 1 the lamp lights.

a) Name logic gate X.

b) Table 6.9 shows all the possible input logic levels to gate X. Copy and complete the table to show the output logic levels of gate X.

| Logic level of left door switch | Logic level of right door switch | Output level of gate |
|---|---|---|
| 0 | 0 | |
| 0 | 1 | |
| 1 | 0 | |
| 1 | 1 | |

**Table 6.9**

**4** A drilling machine works only when the operator presses a switch and a safety guard is correctly in position. Part of the electronic circuit for the drilling machine is shown in figure 6.26.

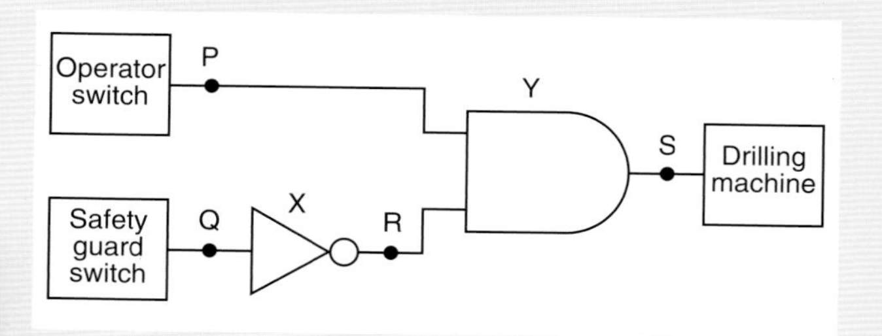


**Figure 6.26**

a) (i) Name logic gate X.

(ii) Name logic gate Y.

b) Table 6.10 shows the possible logic levels of inputs P and Q. Copy and complete the table to show the logic levels at R and S.

| P | Q | R | S |
|---|---|---|---|
| 0 | 0 | | |
| 0 | 1 | | |
| 1 | 0 | | |
| 1 | 1 | | |

**Table 6.10**

A

acceleration 127
aerials
 radio 3
 satellite 14
 television 8
aircraft
 engine/drag force 121
 sky diving 120
 sound levels 99
air friction 117
airports, scanners in 72
air resistance force 119
ammeters, and electricity 30, 31, 32
amperes (A) 30, 31, 32
 fuses (three-pin plugs) 46
amplifiers
 process sub-systems 140
 radios 4
 sound 101–2
 television 8
 voltage gain 103–4
analogue signals 148
AND-gate 149–50
angle of incidence 16
argon lasers 59, 60
arthritis 78

B

baby alarms 140
Baird, John Logie 7
balls (sports)
 bouncing 134–6
 forces 111, 131
 impacts 131
banknotes, forged 80
baseball, and impacts 131
bathrooms 49
batteries
 circuit symbols 29
 circuit testing 51
 positive/negative terminals 30
 voltage 33
Bell, Alexander Graham 21
blindness 59, 60
bones, X-ray use 70
bouncing 134–6
branches (electrical paths) 30, 35
British Broadcasting Corporation (BBC) 7
bronchoscopes 63
buzzers 21, 144

C

cameras
 gamma 75
 heat seeking 78
 speed 127
cancer
 endoscopes, viewing with 64
 gamma radiation 74
 infrared radiation 78
 laser treatment 58, 59
 skin 81, 82
carbon dioxide lasers 59
cars
 air friction, reducing 117
 balanced forces 119
 brakes 114, 115
 collisions 129–31
 engines 116
 racing 117
 speed, measuring 124, 126
 streamlining 117
cartridge fuses 46
cataracts 83
 operations (laser treatment) 60
cells/cell phones *see* mobile phones
circuit breakers 45
circuits, electrical
 domestic 44
 parallel 32, 34–5
 series, measurement of current in 31
 symbols 29, 45, 150
 types 30
 voltage measurement 33
circuit testers 51
cladding glass (optical fibres) 17, 63
Clarke, Arthur C 11
cold light source 63
collisions 129–31
colour TV tube 9
communication/communication systems, satellites 12,14
compact discs 58, 104
computers
 e-mails 24
 instantaneous speed, measuring 125
 internet 20, 22
concave lenses 65, 68
conductors 29, 49
consumer units 44, 45
contact lenses 68
converging lenses 65, 66, 67
convex lenses 65, 67
copper wires/cables 17, 18
cornea 61, 66
credit cards, hologram use 61
current, electric
 defined 28
 fuses 46
 mains electricity 44
 measurement 30, 31
 resistance 39
curved reflectors (dish aerials) 14
cyclists 118

D

DABS (Digital Audio Broadcasting System) 5
d.c. power supply 145
decibels 98
declared value (mains voltage) 44
decoders 4, 8
dentistry, ultrasound use 98
diabetes 59
Digital Audio Broadcasting System (DABS) 5
digital signals 148
dish aerials 14
diverging lenses 65, 68
double insulation 48
drag force 121
DVD writers 58

E

ear
  damage to 98, 99
  hearing, range of 97–8
Early Bird (geostationary satellite) 14
earplugs/earmuffs 99
Earth
  gravity 112
  satellites 11, 13, 14
  UV radiation 81
Earth wire 46, 47, 48
echo method 94
electrical system, defined 140
electrical telegraph systems 20
electric drills 48
electricity
  appliances 35
  circuits *see* circuits, electrical
    conductors 29, 49
  current *see* current, electric
    defined 28
  energy conversions 28, 35
    insulators 29
  mains *see* mains electricity
    television receivers 8
  voltage *see* voltage
electricity bills, calculating 50
electricity meters 44
electric motors 144
electric shock/electrocution 49, 50
electron gun 9
electronic scanning
  (television) 7
electronic system
  as block diagram 141
  signal use 148
electrons
  currents, electric 28, 30
  television picture tubes 9
  voltage 33
element, defined 43
e-mails 24
encryption 20
endoscopes 63, 64
energy conversions 28, 35
equator 13
eye
  cataracts 60, 83
  cornea 61
  functions (sight, description of) 66–7
  laser treatment 59–60
  long sight 67, 68
  short sight 67–9

F

facsimile (fax) machines 24
factories, sound levels 99
fault finding 52
fibre optics, use in medicine 63–4
filaments 43
fire risk 46
fires, electric 43
flexes (flexible cord) 46, 47
  safety factors 50
food mixers 50
forces
  balanced 118–20
  ball, acting on 111, 131
  calculating 113
  car brakes 114, 115
  frictional 114
  reducing friction 116
  measuring 111–12
  moving objects 131
  streamlining 117
  unbalanced 121
  variety 110
  weight 112
frequencies/frequency bands
  mobile phones 23
  octaves 89
  radio 3, 5
  satellites 14
  sound 87
  television 6
friction 114
  reducing 116, 117
fuse boxes 44
fuses 45, 46–7

G

gamma radiation 74, 75, 76
gases 97, 117
geostationary satellite 13–14
glasses (spectacles) 68
gravity 112, 136

H

hairdryers 48
hearing, range of 97–8
heat seeking cameras 78
helium neon lasers 58
helmets 99
Hertz, Heinrich 2
hertz (Hz) 3, 89, 97
holograms 61
house wiring 44

I

incident rays 62
industry
  ultrasound use 98
  X-rays in 71
infrared radiation 78–9
input devices 142–4
input sub-systems 140
instantaneous speed 124
  measuring 125
insulators 29
internet 20, 22
inverters (NOT-gates) 149

K

kettles 43, 50
keyhole surgery 63
kilograms (mass measurement) 113
kinetic energy 144

L

lamps
  circuit symbols 29
  circuit testers 51
  circuit types 30
  electric currents 28, 31
  energy conversions 35
  flashing 20
  fluorescent 80
  heat 79
  light emitted from 58, 62
  output devices 145
  power, electrical 44
  voltage 33, 34
lasers
  meaning 58
  optical fibres 18
  uses 58–60

laws
  Ohm's law 40
  reflection 16
LCD (liquid crystal display) screens 7
LDR (light-dependent resistor) 143
LED (light-emitting diode) 18, 145
lenses
  long sight 67
  short sight 67–8
  sight, description of 66
  types 65
Leonardo da Vinci 58
light
  bending of (in sight) 66, 67, 68
  lamps, emitted from 58, 62
  LDR (light-dependent resistor) 143
  LED (light-emitting diode) 18, 145
  OLED (Organic Light Emitting Diodes) 10
  rays of 62
  reflection of 16, 62
  sending messages with 17–18
Light Amplification by Stimulated Emission of Radiation *see* lasers
light gate
  instantaneous speed, measuring 125
liquids 96,117
listening (radios) 4
live wire 46, 47, 48
logic 0s and 1s 104, 150
logic gates, digital 148–56
  AND-gate 149–50
  defined 149
  NOT-gate 149
  OR-gate 150
  signals 148
long sight 67, 68
loose connections 51
loudspeakers
  amplified sounds 102
  human hearing, testing 97
  output devices 140, 144
  radios 4
  telephones 21
  television 8

M

magnetism 104
mains electricity
  bills, calculating 50
  circuit breakers 45
  circuit testers 51
  current 44
  dangerous situations 49–50
  double insulation 48
  fault finding 52
  flexes 46
  fuses 45, 46–7
  house wiring 44
  plugs 46, 47–8
  power 43, 44
Marconi 2
mass (weight) 113
measurements
  electric currents 30, 31
  forces 111–12
  rays of light 16
  resistance 39, 40
  speed 123, 125, 126
    of sound 93–4
  sound levels 98
  voltage 33–4
  weight (own) 113
medicine
  fibre optics, use in 63–4
  gamma radiation 75
  infrared radiation in 78–9
  laser use 58, 59–60
  ultrasound use 98
  X-rays 70–2
melanin 83
melanoma, malignant 81, 82
metal fatigue 98
microphones
  amplified sound 102
  input devices 140, 142
  optical fibres 18
  sound waves 88
  telephones 21
  *see also* transmitters
mobile phones 22–4
  text messaging 20
Moon 11
Morse, Samuel 20
motion *see* moving objects
motors, electric 144
moving objects
  bouncing 134–6
  collisions 129–31
  forces 131
  height, and bouncing 134–6
  impacts 131
  range of throw 132–4
musical instruments 87
myopia (short sight) 67–8

N

neutral wire 45, 47
Newton balance 111, 112
newtons 113
Newton, Sir Isaac 112
noise pollution 99
normal (light reflection) 62
NOT-gate 149

O

octaves 89
ohmmeters 40, 143
ohms 39
Ohm's law 40
OLED (Organic Light Emitting Diodes) 10
open circuits 52
optical fibres 16–19
  composition 17
  detection 18
  law of reflection 16
  light reflection 16
  link trunk systems 18
  medicine, use in 63–4
  message sending 17–18
  reasons for use 17
  transmission 18
Organic Light Emitting Diodes (OLED) 10
OR-gate 150
oscilloscopes
  amplifiers 4
  analogue/digital signals 148
  input devices 142
  'seeing' on 88

telephones 21, 22
output devices 144–5
output sub-systems 140

P

parachutes 120
parallel circuits 32
voltage in 34–5
particles 96
period (satellites) 13
photodiodes 18
physiotherapy 79
piano notes 89
picture tube (television) 9
pixels 10
plane mirrors 62
planets 112
*see also* Earth
plugs 46
three-pin, wiring 46, 47–8
pneumatic drills 99
port wine stains, laser treatment 60
power, mains electricity 43, 44
process sub-systems 140

R

racing cars 117
radiation
gamma 74, 75, 76
infrared 78–9
safety factors 76
time and 76
tracing and 75
ultraviolet (UV) 79–83
radio 2–6
aerials 3
amplifiers 4, 102
decoders 4
frequency bands 5, 14
method of operation 2–3
reception 3
signals 3
transmitters 2
tuners 4
radiography 71
radio stations 5
rating plates 48
ray boxes 16
rebound 135, 136
receivers
radio 2
telephone 18, 21
television 7–8
reflected rays 62
reflection hologram 61
reflection of light 16, 62
refraction 68
resistance
circuit testers 51
measurement 39, 40
resistors 39–40, 51
retina
laser use, eye problems 59, 60
long sight 67
seeing, description of 66, 67
short sight 67, 68

S

safety factors
ear protection 99
flexes 46, 60
plugs, wiring 47
radioactivity 76
satellites 11–15
aerials 14
communication 12
dishes 13, 14
geostationary 13–14
meaning 11
signals, receiving from 13
uses for 11
scales, weight measurement 113
scalpel, laser as 59
scanners, in airports 72
service cables 44
ships, and balanced forces 119
short sight 67–8
sigators 4
signal generators 97
signals
amplified sound 101, 102, 104
analogue and digital 148
electrical, in sight 67
radio 3
telephones 20, 21
television 6, 8
SIM cards (mobile phones) 22
skin cancers 81, 82
skis, waxing 116
sky divers 120
smoke signals 20
solids
sound waves 96, 97
ultrasound use 98
sonometers 87
sound
amplified 101–2
meaning 86
measuring levels of 98–9
noise pollution 99
octaves 89
speech 104
speed of 92–5
ultrasound 97, 98
using 96–9
vibrations 86, 87, 96
sound level meters 98, 99
sound waves 86–91
spectacles 68
speech 104
speed
acceleration 127
average 124
defined 123
instantaneous 124, 125
measuring 123, 125, 126
of sound 92–5
speed cameras 127
speed guns 126
sports
bouncing a ball 134–6
and impacts 131
range of throw 132–4
skis, waxing 116
sports injuries 79
springs, force measurement 111–12
stethoscopes 97
stopwatches 93, 123
VASCAR speed system 126
streamlining 117
stringed instruments 87
sun 62
sun beds 81
suntan lotions 83
swimwear 118
switches 143–4
symbols, electrical circuits 29, 45, 150

### T


tape cassette, compact disc compared 104
tattoo removal, lasers 60
telecommunications *see* optical fibres; radio; satellites;
telephones; television
telephones 20–4
  fax machines 24
  history 21
  mobiles 20, 22–4
  optical fibre use 18
  oscilloscope patterns 22
  receivers 18, 21
  signals 20, 21
  transmitters 21
television 7–10
  amplifiers 102
  optical fibre use 18
  receivers 7–8
  signals 6, 8
  transmitters 6, 8
  tubes 8, 9
tennis, and impacts 131
terminals, battery 30
text messaging 20
thermistors 142, 143
thermograms 78
thunder and lightning 92
toasters 48
tracing, and gamma radiation 75
transmitters
  optical fibres 18
  radio 2
  satellites 14
  telephones 21
  television 6, 8
  *see also* microphones
transport, means of *see* aircraft; cars; ships
triangles (colour TVs) 9
truth tables 149, 150
tubes, television 8, 9
tumours *see* cancer
tuners 4, 8
tyre tracks 130


### U


ultrasound 97
  use in medicine and industry 98
ultraviolet radiation (UV) 79–83
UVA radiation 81, 82
UVB radiation 81, 82
UVC radiation 81


### V


vacuum cleaners 46, 48, 144
VASCAR (speed system) 126
vehicles *see* aircraft; cars; ships
vibrations, sound 86, 87, 96
video phones 22
voltage
  amplifiers 103–4
  mains electricity 44
  measurements 33–4
  in parallel circuits 34–5
voltmeters 33, 34, 35
volts (V) 33


### W


washing machines 144
water, and electricity conduction 49
water boilers 72
watts (W) 43, 46
weight 112
  calculating 113
wind instruments 87


### X


X-rays 70–2